Remember Me

Remember Me

A LIVELY TOUR OF THE
NEW AMERICAN WAY OF DEATH

LISA TAKEUCHI CULLEN

Collins
An Imprint of HarperCollinsPublishers

HarperCollins books may be purchased for educational, business, or sales promotional use. For information, please write: Special Markets Department, HarperCollins Publishers, 10 East 53rd Street, New York, NY 10022.

FIRST COLLINS PAPERBACK EDITION PUBLISHED 2007

Photographs by the author, unless otherwise stated.

The Library of Congress has catalogued the hardcover edition as follows:

Cullen, Lisa Takeuchi.
  Remember me : a lively tour of the new American way of death / Lisa Takeuchi Cullen.
    p.   cm.
  ISBN-13: 978-0-06-076683-2
  ISBN-10: 0-06-076683-2
    1. Funeral rites and ceremonies—United States.   2. United States—Social life and customs.   I. Title.

  GT3205.C85 2006
  393'.9—dc22

2006040273

  ISBN 978-0-06-076684-9 (pbk.)

07 08 09 10 11 ❖/RRD 10 9 8 7 6 5 4 3 2 1

For my father, Thomas J. Reilly,
who loves a good party.

Contents

CONTENTS

The Dead Beat

A REPORTER'S CHALLENGE: FINDING THE FUN IN FUNERALS

Mika and I check out a vintage hearse at the Frozen Dead Guy Days festival in Nederland, Colorado.

What do you wear to crash a funeral?

This is a dilemma I can honestly say I never anticipated. It's 2004, I have just had a baby, and I am writing a book about death. To report this book, I will have to attend some funerals uninvited, in the process of which I would prefer not to commit some unforgivable funeral-fashion Don't.

I stand before my closet. There is an abundance of black in my wardrobe, not because I am morbid—which I swear I'm not, subject of first book notwithstanding—but because I am a woman who works in New York. I nix my favorite black pantsuit for fear it screams corporate. Here is a classic black shift that used to look great on me before a small person came out of my abdomen. I settle on black pants and a black sweater, both snug in weird places but relatively free of dog hair and spit-up.

The service in question is not precisely a funeral but a memorial service for the recently bereaved, held at the Ippolito-Stellato Funeral Home in Lyndhurst, New Jersey. "Aftercare" is a buzzword in the funeral industry; where once a funeral home's role ended with the closing of the grave, today it continues with gatherings like this one, meant to share and thus ease mourners' grief. Inside the funeral home, the lighting is warm, the plum-colored carpeting is plush, and the furniture is elegantly upholstered. There is a plump Christmas tree in the lobby accessorized with gold ribbons. Images of the deceased waft onto a TV screen in a slide show set to music. On a table in another room await ornaments for each family to place on the tree in memory of their loved one. The theme this year, we are told, is candles.

In the main viewing room with the one hundred or so guests, I listen to a speech by Lou Stellato, the funeral director, about loss and ritual and hope. I bow my head in prayer. I consider, as requested, the metaphoric symbolism of the candle.

In between, I sneak peeks at the real mourners, scanning grief-weary faces, wondering about the ones they lost and what kind of casket they were buried in.

And I ask myself, not for the last time: How did I find myself here?

My own tour of the new American way of death began in 2003, when I was assigned to write an article for *Time* magazine, where I work, about funeral trends among baby boomers. The assignment had little to do with Jessica Mitford and her classic 1963 exposé of the funeral industry, *The American Way of Death*. My editor was instead interested in the personalized and often wacky ways Americans were reinventing the rites and rituals of death. NASCAR coffins! "Green" burials! Diamonds made of human remains!

Here's a perhaps unsurprising confession. Sometimes we journalists fashion trends out of wishful thinking. The magic number is three: Three examples of anything do a trend prove. But this was more than that. As the seventy-six million or so people born between 1946 and 1964 began to hit sixty, they were confronting death en masse in the loss of their parents or each other. And one after another of them, it seemed, was spitting on the status quo.

On the face of it, this, too, should surprise nobody. After all, boomers stand accused of bulldozing most cultural norms, from sex to music to hairstyles. Death merely came next on the to-do list.

But consider the scale of this makeover. Religion, for one, still dictates most funeral and burial rites in this country, as around the world. Yes, boomers, like the rest of the population, are increasingly secular; between 1990 and 2001, the number of adults identifying themselves as "nonreligious" nearly doubled, from 8 percent to 14.1 percent, according to a national survey of religious identification conducted by the City University of New York. However, refusing to cede even the last rites to God spoke to a degree of independent-mindedness I had not suspected.

For another thing, the status quo has a well-armed enforcer: the funeral industry. Americans' expectations of what happens at death have been shaped, when not by religion, by funeral directors. With up to $20 billion* of annual revenue at stake, you can bet the industry is ready to rumble.

The rumble, in fact, is in its umpteenth round. Jessica Mitford's book triggered a national outcry against what she defined as the funeral industry's controlling and predatory practices. It wasn't until two decades later, 1984, that the Federal Trade Commission finally implemented its Funeral Rule, a set of laws meant to regulate the industry. Have they worked? Nope, said Mitford in her 1998, posthumously published follow-up, *The American Way*

---

* Estimates of the size of the U.S. death care industry vary, from $11 billion (National Funeral Directors Association, though this does not include burial costs) to $20 billion (various news reports). The NFDA says the average cost of a funeral as of July 2004 is at $6,500; burials averaged around $3,000 and cremations $1,000 (Cremation Association of North America). Of course, not everyone who dies gets a funeral and burial, but the industry organizations do call these the *average* costs. At 2.3 million deaths a year, with 25 percent cremated in 2003 (the latest number available from CANA), I figure this adds up to $20 billion—rounded down.

of Death Revisited. It would surprise her not at all that in 2005, consumer advocates slapped the country's biggest funeral home chains with a major class action suit accusing them of conspiring to set prices.

Maybe the lawsuit will succeed in transforming the old American way of death; maybe it won't. Me, I'm betting on the boomers. Currently, about 2.3 million people die every year in the United States, according to the National Center for Health Statistics. By 2040, as boomers move onward and upward, the number of deaths will double. Forty years have passed since the so-called revolution of the death business began, and it's finally becoming clear that the only force powerful enough to change this intransigent industry is the force of the market. My money's on the seventy-six million people who'll want to do death their way.

Let me say that I am not Jessica Mitford. This book does not expose the foibles of the funeral industry, except insofar as their attempts to keep up with the times. (This involves polka dots and pigeons. See "Confessions of a Funeral Planner.")

Neither does it explore religious rites or the history of funerals. I did not attend lectures on thanatology or cogitate (much) on what it means to die.

What I wanted to know was this: What kind of person turns a loved one into jewelry? What's it like to watch an artificial reef mixed with the cremated remains of your parent sink to the bottom of the sea? How exactly is a modern mummy made? Where would I find a festival celebrating a frozen corpse? What's the proper etiquette at a funeral involving animal sacrifice? Who would become a funeral director today—and why?

In other words: What's it like to be a consumer shopping for after-death options today?

I wanted to meet these people, the consumers as well as the merchants I secretly called the end-trepreneurs. I wanted to scatter ashes from an airplane, hang out at a modern mortuary school, touch the casket shaped like a lobster. I wanted to tour the earth-friendly cemetery, see a plastinated body, make a midnight visit to a frozen dead guy in a Tuff Shed.

So I set out on a tour of what I saw as the new American way of death.

Death happened during the time I reported this book. Death happened a lot. At times, I felt the front page read like the obituaries.

Ronald Reagan's funeral in June 2004 led the news for days, muting both red states and blue with its pageantry and emotion. At the close of the year, we learned with horror of the tsunami in Southeast Asia that swept away thousands of lives, leaving little in its wake to memorialize. The following April, I along with the world watched the three-hour televised funeral of Pope John Paul II, mesmerized by the Gregorian chant, the kyrie, the parade of cardinals around the cypress-wood coffin.

Around the same time, the nation argued over the impending death of Terri Schiavo, the brain-damaged Florida woman over whose life her parents and husband fought so bitterly. If anything positive came of their public battle, it was perhaps a wider awareness of the need for living wills, not to mention a document stating our desired mode of disposition. (Terri's parents wanted her body buried in a cemetery, according to news reports; her husband chose cremation; Terri's own wishes, we'll never know.)

Of course, the war in Iraq ensured there was never a shortage of death in the headlines during the course of my reporting. In October 2005, the death toll of U.S. soldiers hit two thousand. Before the war began, my generation had only vague notions of the military funeral. Today, we cannot escape the images of wives and mothers hugging folded flags to their broken hearts.

And then there was New Orleans. A few years ago, I had toured the cemeteries with their ornate, aboveground crypts. I chose not to return while reporting this book, even though the funeral tradition there is perhaps the best known of all American death rites and the ancestor of the celebratory rituals of today. If I had, I might have witnessed a brass-and-wind band's dirge escorting a family from home to church. I might have heard the "cutting loose" of joyous sound after the burial, stood among the crowds of jubilant mourners. I could have visited Congo Park, where slaves are said to have pounded drums to send the spirits of their dead back to their ancestral land. I might even have caught the jazz funeral's modern-day descendant, the hip-hop funeral, like that of rapper Soulja Slim, whose death in 2003 drew thousands to the streets.

But I missed my chance.

At the close of summer 2005, Hurricane Katrina ripped through Louisiana, tipping the contents of Lake Pontchartrain into the basin-shaped city. New Orleans had drowned. But there would be no joyous-sad parade down the ruined streets for the 972 residents who were lost, no brass band or quick-step dancers to mourn and celebrate the dead.

Others stepped in. The romance of the New Orleans funeral so enamors the rest of the country that we jumped at the chance to copy it. Semiorganized or entirely impromptu "funerals" in the wake of the disaster were reported in New York City, Atlanta, and countless smaller towns. So when I saw a listing for a "jazz funeral for New Orleans" in my local New Jersey newspaper, I felt compelled to check it out.

By the time I arrived at Cooper's Pond in Bergenfield, New Jersey, a couple of dozen people were gathering by the pavilion. Five musicians were warming up on their instruments—two trumpets, two trombones, and a clarinet—near a monument to a more local disaster, the 9/11 attacks. (Bergenfield, like most of the towns in this suburb of New York City, lost residents to that tragedy.) Votive candles lined the walk.

The broad, bearded brass player wearing a captain's hat that read "Bone Tone" was Rabbi David Bockman. The gathered crowd were members of the Congregation Beth Israel. The rabbi, who had moved here recently after six years in New Orleans, organized the funeral march to share his love and memories of his former home. After the march, there'd be a pre–Rosh Hashanah *Selichot* service back at the temple.

The band struck up the first notes of "Just a Closer Walk with Thee," and the congregation began to hum along. To most, the lyrics were clearly unfamiliar; spirituals, after all, are rarely sung in temple. A leather-jacketed man standing near me knew the words, and his voice soared above the others':

When my feeble life is o'er,
Time for me will be no more;
Guide me gently, safely o'er
To Thy kingdom shore, to Thy shore.

I marveled at the incongruity of the New Orleans jazz funeral, a rite descended from African slave rituals, celebrated by white northeastern Jews. I do not know how African American New Orleanians would feel about the rest of the country trying on their ancient funeral rites. But I found it touchingly American, this eagerness to embrace what moves us in other cultures and make it our own.

———

Some might think my tour ghoulish or macabre. I am neither of those things.

I do not harbor a death fetish, unlike some of the people I met along the journey. I have never worn the "I Put the Fun in Funerals" T-shirt that Mark Chiavaroli, proprietor of CityMorgueGiftShop.com, gave me as a souvenir when we met in Los Angeles. On the other hand, I do not suffer from a death phobia. I do not regularly check DeathClock.com to see how many seconds I have left (for the record, my "personal day of death" is July 7, 2061).

I felt confident I would take a reporter's detached approach to the subject of changing death rites, a journalistic interest in their social, cultural, and business implications.

Until I had a child.

It seemed like a good idea at the time. I would have a baby, then write a book during maternity leave. The book advance would help me take a longish leave, and the leave would allow for lots of in-person, on-location reporting.

With the baby.

I concede. It never seemed like a good idea to anybody but me.

Three months after I gave birth, still stupid with postnatal hangover, I awoke to the realization that the national funeral directors' convention was about to start. I threw some Huggies into a bag, packed Mika into a Björn, and elbowed our way into the last seat on a flight to Nashville.

At best, I figured navigating an undertakers' convention with a baby in a stroller would brand me a freak. As it turned out, we were hardly alone. Funeral homes, after all, are a family business, and entire generations make the annual trip. I stopped other new moms for impromptu kaffeeklatsches by the casket displays, trading notes on the best lavatory lounges for nursing and diaper changing (Washington wing, past the Roy Acuff gun collection). What's more, my googly-eyed sidekick seemed to earn me unwarranted trust from funeral directors, many of whom are leery of press.

Of course, not everyone was so accepting. Once, on an airplane, I sat next to a woman in a business suit who spoke cheerily to me about motherhood. The book in her lap was Dale Carnegie's *How to Win Friends & Influence People*. She was on chapter 2, "How to Make People Like You." I pulled out my own reading material, a recent copy of *American Funeral Director* magazine.

"Are you a funeral director?" she asked.

"Oh, no," I said, adjusting Mika's car seat.

Her smile sagged. Covering the dead beat as a new mom was not always conducive to making people like me.

On my many trips around the country and twice to Japan, I simply took the baby along. For the most part, this worked out. Mika slept through the coffin race at the Frozen Dead Guy Days festival in Nederland, Colorado. She played in the tall grass at a natural-burial cemetery in Westminster, South Carolina. She peered at the death mask a woman had made of her best friend in Occidental, California.

And I spent those nights watching her breathe. I spent those nights watching for the rise and fall of her chest. Watching to make sure she did not grow still.

Giving birth made me certain of death.

Death—until then, unsurprisingly for someone my age, a hazy and abstract notion—became real, possible. Death, I realized, could happen to her. It could happen to me, which now mattered because of her.

I could not stomach even thinking about children's funerals. I knew they happened because some companies deal exclusively in infant caskets. I knew these companies existed because I speed-walked past them at the Nashville convention, trotted right past their displays of tiny little coffins lined with pink or blue gingham, their racks of white outfits that look like baptismal clothing but are meant for the underground. These booths at the convention were always empty. No one, not even the steeliest funeral director, wants to hang around that kind of karma.

I figured out quickly that new parents instinctively reject death. How can we not? Mother Nature made us that way to ensure the survival of our little ones. My body coursed with hormones that made my brain rebel against the kind of information I was trying to pack in. Like which dead celebrity earns the most (Elvis, with $40 million in 2004). Like the top ten list of songs British people want played at their funerals (number one: Queen, "The Show Must Go On"). Like a routine Jerry Seinfeld performed on *Letterman* about cremation: "It's kind of like covering up a crime—burn the body, scatter the ashes around. As far as anyone is concerned, the whole thing never happened."

I read a lot of obituaries. I read them every morning, over breakfast, while Mika spooned yogurt into her hair.

In the weddings section of the Sunday *New York Times,* there is a regular feature called "Vows." It tells the story of a couple, beginning with how they met cute (haggling over coffee beans at a street fair in Instanbul) to the rough spots (he discovers she can't stand cilantro, a cooking ingredient without which his chicken curry is incomplete) to a sumptuous description of the glorious day (peonies and ponies and a lesbian Hindu officiant).

As I read countless death notices looking for interesting funerals to crash, I couldn't help but wish they read more like those articles. Instead of accumulated accomplishments accompanied by a dated head shot, I wished they gave a sense of who these people were, a taste of the life they lived, a hint of the fabulous celebration planned in their honor.

• A NOTE ON SOURCES •

Which brings me to the logistical difficulties of covering the dead beat for a reporter with a baby.

One thing I had failed to think through was that death cannot be scheduled. Subsequently, one cannot plan a trip in advance for a funeral and burial. I did try. At one point, the proprietor of the cemetery in South Carolina gave me regular updates on the state of two prospective customers who were then in hospice. Alas, my diaper-wearing assistant required arrangements of a sort that mostly ruled out last-minute travel.

As a result, I sometimes had to play a sort of funeral CSI, trolling the premises, poring over photographs and videos, pressing organizers or family and friends to describe and even reenact the proceedings. The events I was not present for, I recounted in the past tense, as the witnesses did.

My tour is not a chronological one. I reported what I could, when I could, mostly over the course of a year (the first year of my baby's life, in fact; this book will have to act as the scrapbook I never made).

The vast majority of my reporting was conducted in person, in face-to-face interviews or occasionally over the telephone. What statistics and research I derived from outside sources such as newspaper articles, academic studies, or market surveys, I attributed within the text for the most part, though sometimes in footnotes.

I read the following books, which contributed greatly to my understanding of death and its rites, its traditions, and its business: Jessica Mitford, *The American Way of Death Revisited;* Mary Roach, *Stiff;* Joan Didion, *The Year of Magical Thinking;* Thomas Lynch, *The Undertaking;* Lisa Carlson, *Caring for the Dead: Your Final Act of Love;* Charlton D. McIlwain, *When Death Goes Pop* and *Death in Black and White;* Gary Laderman, *Rest in Peace: A Cultural History of Death and the Funeral Home in Twentieth-Century America;* Anne Fadiman, *The Spirit Catches You and You Fall Down: A Hmong Child, Her American Doctors, and the Collision of Two Cultures;* Evelyn Waugh, *The Loved One;* C. S. Lewis, *A Grief Observed;* and Timothy Leary, *Designer Dying.*

The editor of *American Funeral Director* kindly put me on his subscription list. I read daily news roundups from FuneralWire.com and received postings from About.com's "Death & Dying" column.

I watched films, including *A Cemetery Special,* a PBS documentary produced by Rick Sebak; *Arlington: Field of Honor,* a National Geographic special shown on PBS; *Grandpa's Still in the Tuff Shed,* by filmmakers Robin, Kathy, and Shelly Beeck; *A Family Undertaking,* by Elizabeth Westrate. I read the transcripts to the excellent 1998 series broadcast on NPR's *All Things Considered* called *The End of Life: Exploring Death in America.*

Mostly I went places and talked to people. I intruded upon families and businesses, pressing for stories and evidence of lives lived and lost. These people had no reason or motive to speak to me other than to preserve the memory of someone they loved.

Remember me. That is all their loved ones asked. And I do. Though some of them I never knew, I remember them well.

Remember Me

Four Funerals and a Wedding

FIVE TAKES ON CELEBRATING A LIFE

*Tango dancers pay tribute to Bette Runge, whose urn
presides in the window of her Washington, D.C., ballroom.*

TOMMY ODOM
1963–2004

Rebecca Love met Tommy Odom ten years ago at a Renaissance fair. She had a booth selling her clay sculptures of women's bodies; he ran the fool's maze. She wore ribbons in her long blond hair; he wore an orange-and-red jester's cap. The bighearted artist and the blue-eyed Texan made a striking and passionate couple, and they stayed together a year before she broke it off. She had learned what those Renaissance women knew: Good-looking jesters with the traveling fair don't always make the best boyfriends.

Their romance unraveled, but Rebecca and Tommy remained close. Rebecca had abandoned her career as a cosmetics executive to focus on her art, settling in a lemon yellow cottage on a brambly acre and a half in California's wine country. He stayed with her whenever the fair took him through, or just when he had nowhere else to stay.

In the summer of 2004, he was preparing to move—permanently this time—into the 1953 metal-sided trailer parked up against Rebecca's cottage. He'd spent the summer clearing brush off her property and starting fix-it projects around the house.

That September, their friend Laura came to visit. She too once dated Tommy. Rebecca, exhausted from a gallery showing in Los Angeles, met up with them at a bar in Occidental. The two had clearly been there awhile.

"I saw them and gave him a look," says Rebecca. "You know, a disapproving one. All I said was, 'I'll see you back at the house.' He was like, 'Okay.' She weaved a bit."

At 1:00 a.m., the phone rang. It was Laura, calling from up the road. Rebecca drove madly in the blackness to where the road bent like a fishing hook. There she saw the ambulance and the medics and Laura lying on the concrete—but no Tommy. A young paramedic approached her. He said, "I'm sorry. He didn't make it."

"No, no, no, that can't be," she replied, stunned. "I just talked to him. You have the wrong person."

With Tommy at the wheel, the car had skidded over the edge of the road, coming to a stop with its nose teetering into the ravine. Tommy shouted for Laura to get out. She did. He lunged out of his window and was almost free when his foot caught on the emergency brake. The car lurched forward. Realizing his mistake, he looked at Laura.

"Bunny," he said, using his friend's nickname. "Oh, no."

Rebecca knew she would bring him home. A funeral parlor was no place for Tommy's body; in any case, he'd left not a dollar. A friend had told her recently of a funeral that was overseen by someone who called herself a death midwife. Jerri Lyons, it turned out, was a leading advocate of home funerals and lived in nearby Sebastopol. Rebecca thought it a sign. Lyons took her van to help her collect Tommy's body from the coroner's office. Together they would prepare it for the funeral and burial.

It would not be pretty. The coroner had performed a standard autopsy, cutting a Y from the shoulders to the groin, slicing the scalp for a piece of brain. He had not cleaned up. Tommy's body arrived at Rebecca's house in a body bag, purple from asphyxiation, crusted with blood.

Lyons set up her massage table in the clearing behind the barn that Rebecca uses as her studio, under the four oak trees where Tommy had liked to sprawl. The two carried his body there on a gurney and unwrapped him. Rebecca used the garden hose to shower off the blood. She soaked rags in rosemary water she'd boiled herself because she'd heard the herb had antiseptic qualities. Gently she wiped his limbs clean.

The September sun was warm, but Rebecca covered Tommy's naked body with a blue blanket. She worked quietly, her silver jewelry jangling and red sundress swishing, attended by a bevy of curious bees.

When his body was clean, Rebecca made Tommy's death mask. She prefers to call it a life mask. As an artist, her specialty is taking plaster molds of naked pregnant women, which she then casts into clay (the license plate on her pickup truck reads LOVECLAY). The clearing where she cleaned Tommy is strewn with her cast-offs, a cemetery of broken torsos and cracked heads. She layered bandages soaked in plaster on his face. "The funny thing is, I was going to cast his face that very week," says Rebecca. "Well, I got to, anyway."

She and Lyons wrapped Tommy's body in white muslin. Rebecca had found a local carpenter to build a plain pine coffin for $484. They placed the coffin on the worktable in her barn, filled it with dry ice, then laid his body in it. She draped a swatch of red silk across his breast as a symbol. Tommy and Rebecca shared a Christian faith, and one night as they sat hashing out the universe by the wood stove in her living room, he had said, "When I die, man, I wanna be covered in the blood of the lamb." Rebecca had complied.

All their friends from the fair turned out for the funeral. Hobbit, Toewinkle, Smiley, and the rest sat in the grass and on foldout chairs in the clearing by the barn. Children chased one another. Hardly anybody wore black. Rebecca, in a green dress that flowed, read from Psalm 23:

Yea, though I walk through the valley of the shadow of death, I will fear no evil; for thou art with me; thy rod and thy staff, they comfort me.

They told stories of Tommy's antics and his sweetness. "He's the kind who'd give his last twenty bucks to a hobo," someone said. "Then he'd turn around and ask you, 'Can I borrow twenty bucks?' " True to form, his friends and family pitched in to pay for the funeral.

The crowd filed into the barn to say good-bye to Tommy. They painted his pine box with rainbows and crucifixes and hearts with wings. "Jesus Loves You," someone wrote. "See you in heaven." "Tommy Blue Skies." "Jester."

A group of them drove an hour south to Marin County, where Tommy would be buried. Rebecca had read in an art magazine a few months earlier of a cemetery that performs "green," or environmentally friendly, burials, in which unembalmed bodies are placed in the ground in biodegradable caskets and without traditional markers. She knew it was what he would want.

At Forever Fernwood cemetery, the friends carried the casket down a steep incline overlooking a valley in a section of the cemetery called Free Spirit. Over the whine of a bagpipe, they lowered the graffiti-covered casket into a hole in the ground, then took up shovels and began to fill it in.

A few months later, Rebecca returns to the site, as she frequently does. There is no granite slab proclaiming its underground occupant, but she knows where her best friend lies. Someone has left a wreath of plastic flowers with a ribbon reading, "Tommy Blue Eyes." The rectangular plot has sunk about a foot, as ground does when there is no vault surrounding a casket. When the ground settles, an oak sapling in a pot nearby will be planted here above his head, and a hunk of fossilized wood will bear his name.

Children's voices ring out from the elementary school in the valley. "There's life out there," says Rebecca. "Life. I know that's what he would want to hear."

✦

BETTE RUNGE
1922–2005

*B*ette Runge had two families. One was born of her sixty-two-year marriage to her doctor husband—children and grandchildren and great-grandchildren who all knew her as "Mama." The other was born of her twenty-year love affair with Argentine tango—students and friends and admirers who knew her as "the teacher." Bette's two families orbited around her like a tango class around its instructor, spinning and kicking and dipping while she choreographed the chaos.

For both families, the center of the universe was a house on Harrison Street in Washington, D.C. The two-story colonial was once coated in a lavender of Bette's liking but is now a white more suited to its genteel neighborhood. Inside, the house is still classic Bette, the walls of her living room flaming red, Indonesian masks and Buddha statues and a musty hookah on display. To its side is a small sunroom whose red tiles she squeezed her students onto until she cajoled her husband into letting her build a real ballroom. She was over seventy then. How the workers ever finished with Bette kidnapping them for tango partners is anyone's guess.

She loved many things and many people, but above all, Bette loved a party. Her two families collided at the birthday bashes she threw for herself every February, her gentle husband shaking hands before he retired upstairs, her grandchildren agog as her dancers slid onto the ballroom floor. Bette would reign in the lavender limelight, filling glasses, matching dance partners.

So it was that her daughter Toni decided both families must gather here in her house for Bette's last party. They wept over her body at the funeral home, just as traditional families do, though in the end the dancers erupted in a tango around her casket. They mourned their loss at St. Ann's Catholic Church, as proper people would, though in the eulogy Toni called her mother "Old Battleaxe" and the program featured a photo of Bette vamping in tight black lace. There is grief, there will still be grief. But when they step through the still-purple door of the house on Harrison Street, Bette demands the celebration begin.

It is a party to do Bette proud. Toni has hired an event planner named Sally Anderson, whose recently launched business, called Tributes, helps families throw celebratory, personalized funerals. Anderson has hired a small army of uniformed waiters to remove jackets and hold out silver trays of sangria. She has employed a topflight caterer to lay out a splendid buffet of South American favorites— beef empanadas and lobster croquettes and a salad of jicama and mango. Bette was a woman who could subsist on dessert, and this was not forgotten: In her honor, on the credenza there are tiered silver platters of pastries and star-shaped chocolate treats.

Were Bette here, the chocolate would have made an appearance, then quickly disappeared into her not-so-secret stash in the bedroom. Were Bette here, she would have reveled in the DVD slide show of her life Anderson has made and is projecting onto the wall above the fireplace. Were Bette here, she would have gloried in the fond recollections of her loved ones, each of whom imparts a Betteism.

"She never changed her hair," says Natalie Marmaras of her grandmother's tight black bob.

"Robert Duvall came for dance lessons," says Mark Secrest, Bette's son-in-law.

"When she turned fifty," says Toni, "she bought herself a purple Barracuda. You know, the muscle car."

"She took us to Lord and Taylor's every week," says Kim Spurry, Toni's daughter. "First we ate at the restaurant there, where everyone knew her. Then she always bought each of us one thing."

"She hoarded chocolate," says Valerie Secrest, a third granddaughter.

Bette's dance family is huddled in the ballroom. Carina Losano sits on the steps descending into the room and gazes across the floor at the bay window, in the center of which sits a pink-and-white urn containing Bette's cremated remains. Losano wears a large rhinestone brooch of two tango dancers that Bette's relatives gave her this morning—Bette's parting gift.

Losano is a fellow instructor who learned of Bette soon after her arrival here from Argentina. "She had a lot of men around her," says Losano. "Handsome, Latino men. Men who always do what Bette says." But in Bette, Losano found kin, an adoptive grandmother. In Losano, Bette found a daily dose of her adoptive homeland. When Losano returned to her home country earlier this year, Bette asked her to bring back some earth. "She loved Argentina," says Losano, tears tumbling, "more than many of us."

Bette Dreisonstock was a daughter of Washington, not Buenos Aires. Born to a U.S. Navy captain and a homemaker, she lived in the country's capital all her life, in a house filled with the treasures her father brought back from his journeys. Climbing the steps of the Children's Hospital to her volunteer job on the polio ward, the exotic beauty with the waist-length black tresses caught the eye of a young doctor, who immediately requested a transfer to her end of the ward. Dr. William Runge married Bette in 1944.

Bette bore three children. She named them Nicole, Toni, and Tad but called them Little Virus, Inot, and Boy. Her husband became Petey, Toni's mispronunciation of Papito. Bette protested housework and dressed in lavender and sometimes served milkshakes for dinner. Raised a Catholic, she at one point declared herself a practitioner of a home-based religion called Rungeism, though her own conversion did not excuse her children from attending Sunday mass.

The couple learned Spanish together so William could build a practice serving the region's growing Hispanic population. His generosity drew impoverished patients, but her personality drew the Latin American glitterati, and before he knew it, William was also the official doctor of the Argentine embassy. It was then that a new Bette awoke. Her tongue wrapped easily around Argentine Spanish. Her enthusiastic but untrained legs wrapped effortlessly around a rough and passionate street dance they call Argentine tango.

The dance begins slowly, with the buzz of a *bandoneon.* The couple stands on the dance floor, erect and motionless. When the South American accordion exhales, the dance begins. The dancers' toes lick the floor in slow circles until their legs intertwine. The woman tilts into her partner like a book on a half-empty shelf. He drags her a few steps, then props her upright. He whips her around like a top until she snaps back, tethered by his hand. In a tight embrace, they spin, feet kicking between each other's legs like furious flamingos.

It was Bette's gift not just to dance, but to instill a lifelong addiction to this dance in others. So it seems only right that they dance at her funeral. After the empañadas and the sangrias and the teary hugs, the dancers arrange themselves on the dance floor. Tino Bastidas, a onetime carpenter from Peru whom Bette transformed into a tango teacher, plants a kiss on her urn. Then the *bandoneon* inhales.

In one room of her house, Bette's family eats dessert. In another room, her other family dances. This may be Bette's idea of heaven.

❖

THOMAS JOHN HAAS
1966–2005

*A*t six feet five inches and 275 pounds, Tom Haas could have appeared as fierce and mighty as the Ku, the Hawaiian gods of war. But his family knew him only as a gentle giant. Raised in Illinois and California, he met his Hawaii-born wife, Judina, and moved to Maui fifteen years ago. There he worked as assistant director of security for the Four Seasons Hotel. He and Judina raised three handsome sons. He trekked the West Maui Mountains and camped out on beaches with friends. Late at night, he penned poetry for his wife.

But perhaps the spirit of the warrior gods coursed in him after all, expressed by his love of knives. He belonged to a fraternity of avid collectors who traded gleaming blades and intricate casings and had recently begun to craft his own. He pulled extra shifts to build up enough vacation for a knife collectors' convention in Las Vegas.

His mother, Jane Glenn Haas, was to meet him there. Jane, sixty-eight, a nationally distributed columnist for the *Orange County Register,* was in Vegas for New Year's with her husband, Bob. Tom's employer had supplied them with complimentary rooms. Calling on his cell phone, Tom told his mother he would be late to their arranged meeting for a showing of the musical *Mamma Mia.* He'd see them at the intermission, he said.

Tom never showed. Jane felt certain her hardworking son had simply crashed in his room. The next morning, the phone remained silent. As Jane began to fret, her husband suggested she call security for a welfare check.

Jane stepped into the hallway as some men approached her room. The longtime reporter in her spotted the badge on one of their belts. The mother in her recognized the look authorities assume when delivering devastating news.

"Is my son dead?" she demanded.

"Yes, ma'am," said an officer.

The autopsy report would take time, but Jane learned that the day before the trip Tom had been involved in a minor car accident, which may have caused internal injuries. A doctor had prescribed pain-killers. Because of his size, "he was a guy who if it said to take two aspirin, he'd take four," says his mother. These things may have contributed to his death. Tom Haas was thirty-eight.

Tom's body was cremated in Las Vegas. "I had to go to the funeral home to make arrangements for his cremation," Jane wrote in her column. "I had to walk past the viewing rooms and the room where the flower arrangements are on display. I had to talk about urns and whether or not I wanted a lock of his hair as a memento."

But Tom Haas belonged to the islands, and that was where he would return and where his wife would orchestrate a service that touched every aspect of the big man's life.

Friends and family filled a Maui funeral home that seated 150. A koa-wood urn of Tom's ashes presided, draped with leis. Framed family photos crowded a table by the door. The program featured a sunny picture of Tom wearing a Hawaiian shirt, above an Irish blessing to reflect his heritage. Friends from California spoke of Tom as a boy. Island friends told of mountain hikes and campouts. Tom's sister read a eulogy written by their brother, delayed by ice storms on the East Coast, calling Tom "a storyteller and a bard." A well-known local singer pulled out his ukulele and sang a love song of his own composition, pulling sunglasses over his face when he began to cry. Afterward, guests feasted on a buffet of Hawaiian-style sushi on monkey-pod platters.

Judina's ancestors held a plot in a Japanese cemetery. However, Tom's urn was too large to fit—though even that urn could not hold all of his remains. Judina keeps the urn in an honored spot in their home. As for the rest of him, she hired a helicopter. She took the ashes of her husband into the air and released them over the Io Valley he so loved to hike.

Jane felt little but numb in the weeks following her son's death. Yet she remembers every detail of the service. "For me it was like dropping into a foreign country, because nothing about the service was familiar to me," she says. "There was no church, because Tom was not religious. Everything was a celebration of him. What was lovely was the genuine emotion of people who were speaking and the great love and respect they had for my son."

No mother expects to attend her child's funeral. Strengthened by the blizzard of letters from sympathetic readers, Jane began contemplating her and her husband's funerals—if only to spare her remaining children the pain of arrangements. She knows what she doesn't want. Her son's service cemented her opposition to the standard rites. "The embalming, the viewing, the dragging into church—it all seems macabre to me," she says. She wants no cemetery burial, either, no thick granite tombstone like her parents'. "I don't go see their grave. That's not how I want to remember them." She wants cremation. "But I do not want to be buried at sea. I've told my daughter that. I went to a burial at sea, and it was very cold, very dark, very wet. I want sunshine and birds and things."

For her husband, she wants music. If she outlives him, she plans to sneak his ashes into the Hollywood Bowl during a Beethoven concert. As for herself, Jane didn't know—until, not long after the service, she attended a Dixieland festival. "I thought, I just found the band I want to hire," she says. "I'd want to rev it up with something like 'The Saints Go Marching In.' The Irish in me always wanted something where everyone would cry uncontrollably. But no. I cried for my son because he wasn't ready to go. Even so, everyone left that service knowing he had a good life. Me, I don't need the organs. I'd rather everyone feel good. Because I've had it good.

"Bring on the band."

(Four months after we spoke in March 2005, Jane Haas lost her husband, Bob, sixty-seven, to complications from cancer and cardiomyopathy. Jane threw a brunch in his honor at the hotel where they were to have celebrated their twenty-fifth wedding anniversary. "I subsequently got his cremains, and I have been disposing of them in various places," she says. Bob's first resting place: an outdoor concert venue in Orange County where Jane attended a Beethoven concert shortly after his death. "It's wonderful that he is where we go and have a good time—not some cemetery where I would go to weep," says Jane. "The only problem was, when I looked on the concert schedule, it said the next week there'd be a concert by Tom Petty and the Heartbreakers. Ah, well. He'll just have to expand his tastes.")

❖

KEN GROUZALIS
1945–2001

FRAN GROUZALIS (*widow*): We had two funerals for Ken. One before he was found and one after.

LOU STELLATO (*funeral director*): We waited for as long as we could. For months, there was nothing. If someone died in Iraq or Vietnam, there was a piece of uniform, something. But if someone goes to work and doesn't come home, there's a tremendous denial. There's nothing that I can do to confirm that death has occurred.

FRAN: It's still hard to believe. We were married thirty-six years. We worked together at the Port Authority. I was a secretary, he was a mailboy. He used to hang out at my desk a lot. He got drafted into the Marine Corps and was stationed in Okinawa during Vietnam. He was there for a short time; he saw Bob Hope—he was happy.

When he came back, he worked in the stockroom at the Port Authority, going to college at night. He got degrees in business and accounting. He got to be supervisor of parking lots at the World Trade Center. He and his whole department were part of the civilian rescue team. There were thirteen men who helped in evacuation. They all had fire certificates. They didn't get the recognition that the police and firemen got, but they did the same thing.

LISA GROUZALIS-JASINSKY (*daughter*): We have a videotape of people getting out from his section. He clearly could have left, but he chose to stay.

FRAN: He called at seven a.m., then again at nine. Said something had happened, he thought a missile. He said not to worry. Then we didn't hear from him again.

LISA: We thought it would be like in 1993, when the bombing occurred. He took that day off. He missed it.

FRAN: Finally, the Port Authority told me there was no hope.

LISA: I'm more of a realist. After a month, I knew. She kept saying, What if . . . They knew that building better than anybody, she says. They could be living if they were trapped in a certain place, she says.

STELLATO: I received the death call on September 20 or so. I recommended we wait before we make arrangements. We didn't know if Ken would be found. Then we talked about what type of person Ken was. We talked about the casket. This is not a standard funeral. We waited, in constant communication.

Two months into it, we said we should have a rite that says his life had been lived. In the Catholic doctrine, you're not supposed to go into a church with an empty casket. That's a vessel for somebody. Because this was such a unique situation, the church and cemetery allowed it.

FRAN: I honestly don't remember very much. I know we chose a casket. My husband was a motorcyclist. He had a Harley, a Corvette. He liked his toys. He always told me he wanted a black-and-chrome casket.

STELLATO: I told her I knew just the thing. One of our manufacturers makes a casket of stainless steel polished to look like chrome. When I saw it at a convention I thought, Wow, this looks like my brother's Harley. I had the catalog. She said, "That's it."

Not a lot of people knew it wasn't Ken in the casket. It didn't seem important.

LISA: We put his motorcycle helmet in it. His favorite glow-in-the-dark boxers. The Harley teddy bear.

FRAN: Messages from us, letters, pictures. Just to know that there was some part of him in there. To know it was not empty.

STELLATO: Waves of people kept coming. Other survivors, neighbors, church people. Hundreds of people. The funeral mass was like a Sunday mass, with virtually not an empty seat. We had a visitation service, and the marines came. They performed their own ceremony.

LISA: We never got to sit once.

FRAN: The motorcycle people rode here in what they call their colors. They wore their leather and their jeans and parked their bikes out front. Some people I'm sure thought it was strange. But I'm sure he would have liked that.

LISA: I wrote a poem I read at the service. About my feelings for him growing up, how he was always there. I managed to get through it. I wanted to do it for him, to make him proud. Having the service was very important to us to have some closure—even without the body.

STELLATO: We waited till after everyone left the mausoleum, then my son and I put the empty casket back into the coach and brought it back here.

And then in January, they found him.

FRAN: In the concourse of Tower Two.

LISA: It meant a lot to have part of him. Otherwise you're always wondering.

FRAN: We had the mementos again in the second service. I thought of actually taking them out of the casket then to take home. But I didn't. It's okay. What I will never, ever part with—he had a collection of T-shirts. Anyplace we ever went, he had to get a Harley-Davidson T-shirt. People bought them for him, too. He had at least a hundred. That, I'll never part with. That is my most treasured possession now.

At home we have a place where we've got everything related to him: a lot of pictures; steel from the World Trade Center in the shape of a cross; oil paintings that people did for us; a little urn that the city of New York gave us with some dirt from ground zero. I had it inscribed.

STELLATO: The concept was, especially for the people not found, that they felt part of them were in ground zero. They *were* ground zero. That's why the urn.

FRAN: We never went to ground zero. We went to the candlelight ceremony in Liberty Park. They read the names of the New Jersey residents who died. Governor McGreevey was there breaking ground, and they let each of us do a shovel of dirt. They let us bring dirt home. I have never been to ground zero, and I don't know if I ever will.

LISA: I used to go all the time, before. I used to go to work with him and help out, do typing or whatever. No, I can't go, either. It's too emotional.

STELLATO: This family, as unfortunate as it is, was fortunate to have that peace of mind to know they had retrieved him. When trying to bridge the gap in the grieving process, it's imperative that you have something to bury, particularly in a tragedy. Not everybody got to have that.

✦

FRANK HUANG WEDS YU-YUN LIAO
1960

I was born in Taichung, a city in the middle of Taiwan. I lived there until 1965, when I came to get a master's in business administration at U.C. Berkeley.

"I met her in high school. Her name was Yu-Yun Liao. I was a senior, and she was a freshman. It is a very interesting story. Do you want to hear it?

"We had a show, like a talent show, in school. She was a very good Chinese folk dancer. I was interested in photography, so I took some picture of her. In fact, the whole show I took quite a few picture.

"I have no intention at all. It was a boy-and-girl mixed school, how do you call it? . . . Yes, coed. At that age, you are very curious about the opposite sex. I show it to a classmate, and he wants to buy this picture of her. His only request is, he says, 'You have got to get her signature on the picture.'

"We are commuter students, so we walk from the railroad station to school. We all walk in a line, like the military. This is the rule. I feel like an outsider, so I have courage. I ask her, 'Can you sign for me?'

"But the boy and the girl are not supposed to be friends, that is another school regulation. She says, 'Stay away, you can get us in trouble.' She took the print, though, and pretty soon it was summer vacation and she never return it. The next semester I say, 'Where is my picture?'

"Ah. She was supposed to be for my friend. But we became girl-friend and boyfriend instead.

"I wouldn't say she's very beautiful. But I like her because she was very smart. I know this by talking to her. She wins a lot of speech contests and was also good in writing. And she dances the folk dance in a costume like in a Chinese opera. She was very good at that.

"I went to college far away, in Taipei. But still we write letters and stay very close. She was in junior year of high school—I was twenty-three, she was nineteen, twenty—when she had like a permanent pain. Her parents, they take her to a local doctor, but the doctor is really no good. He sends everybody fourteen kilometers away to Taichung to take out the appendix. It turns out she was allergic to penicillin. So after the operation, they give her penicillin to control the inflammation. That did it. That night, she pass away.

"I got a phone call. In fact, the place that I stay at the time, my uncle's house, doesn't even have a phone. So they call the school. One of my classmate knew where I live. He came to tell me. First he says, 'Your aunt pass away.' He doesn't know who it is; he only knew the name. When he says the name, it is really a shock.

"Immediately I took a train and rushed down. I went to their house in the country. I walk in there. I still couldn't believe. And then I see they put her picture in the altar. They tell me the story.

"I didn't even see her last face. Her father says her skin was blue like a poison.

"I went with her parents to pick up the body. But already it was cremated. The doctor tells me she had a brain disease and it is infectious, so that is why they have to cremate. I think the doctor is trying to hide some mistake. Her parents are farmers, and they don't have much knowledge. Someone said you need to have an autopsy, but the parents' traditional feelings say they don't want the body to be cut up. By the time I hear about it, it's too late.

"Later they ask me to marry her.

"It is pretty common in Taiwan that people marry someone who is dead. You can say it is superstition, but don't forget the background in Taiwan is Buddhism. They believe in reincarnation, they believe in hell and heaven. People think if their daughter die young, it leaves unfinished business in their cycle of reincarnation. Even though you're dead, the spirit is still there and you don't want to be alone. So you marry the spirit of passed-away people.

"We had not talked about marriage. We love each other very much, but we are still so young. After she die, of course it has broken my heart. Her parents want the daughter to get married. So I feel it is something I can do.

"In Taiwan, many people believe in fortune-telling. Everybody in my family has a fortune book that tells what will happen from the day you are born until the day you die. I have one, yes. And it is very interesting. There's a poem for every year, and sometimes they don't say specifically, so you almost have to guess. I didn't understand it until a fortune-teller read it. According to the fortune-teller, I had to do this or my living spouse, when I get one, would not have good health.

"One thing I really want to explain. I don't believe so much in this kind of custom or superstition. When I agree to marry her, I say, 'This is the least thing I can do for her. This thing is her parents' only worry. I will do it to make her parents feel comfortable.'

"It was held at her house, a year after she die. You go over and propose to family. We had to call a matchmaker to be part of the ceremony. We give her family money for engagement, then her family give money back as dowry. Everything is symbolic.

"After I graduate and do my military service, I start working in the finance field. Then I meet my wife, Deborah. We are married since 1965. We moved to the U.S. so I could go to graduate school. I work for many years now in finance at IBM, and we live for many years in Danbury, Connicticut. We have four children, three daughters and one son.

"My wife has nothing to say about the wedding. She says it was before her. She believes it is a blessing to herself because the fortune says otherwise she will not be healthy. My children, when I told them, they had not much reaction. It happened long before they were born.

"It sort of changed my philosophy to life. When you were young, you thought life will never change. Young and stupid, I guess. But you start realizing that life is . . . well, we have a saying: The sky can have a cloud all of the sudden, and likewise, something can happen to your life all of the sudden. I guess for me it was reality.

"I realized life is not always what you expect."

❖

Confessions of a Funeral Planner

THE CASE OF THE TRAGIC DOVE RELEASE
AND MORE MODERN-DAY MISHAPS

*At a workshop in Avon, Connecticut, funeral directors learn the ABCs of
personalizing funerals—this one for a schoolteacher.*

Funeral directors don't get out much. Services begin in the morning. Removals happen at 2:00 a.m. In between, there's embalming
and faxing and vacuuming. Living as many do a floor above the shop, they're
never really off the job.

So the annual occurrence that is the National Funeral Directors Association convention is a big, big deal. Between three thousand and four thousand undertakers converge upon Chicago or New Orleans or Orlando to
schmooze and take workshops and check out the latest in death merchan-

dise. The year's vacation is planned around those few days, often for the several generations who together work the business, which explains the crowd of grandmas cooing over baby strollers near the hearse expo.

The 2004 convention is held in Nashville, Tennessee, in an architectural freak show called the Gaylord Opryland. It's a glass-enclosed universe with rain forests and streetlamped boulevards and restaurants housed in plantation-style buildings. At the front desk, the clerk prints out a maze with a squiggly, computer-drawn line, which turns out to be a map showing the way from the lobby to our room.

On the way there, I push Mika's stroller past a gazebo and across a bridge, where we halt a minute to stare at a touring riverboat. The passengers, having not much else to do, stare back. Some wave at the baby. It's weird. The place is *Gone with the Wind* meets Pauly Shore's *Bio-Dome*.

The convention center's past-meets-future theme is fitting. The funeral business is making such an introduction itself, as decades of unrivaled, unquestioned, set-in-granite tradition meets this sonic force called the baby boom. Undertakers can't ignore the rumble of change, not when its own trade association, the National Funeral Directors Association, titles this year's convention "Rhythm of the Future." Not when the NFDA itself says 62 percent of boomers want "personalized" funerals. Not when *American Funeral Director* magazine finds in its 2004 consumer survey that 71 percent do not want a traditional funeral, with 14 percent specifically requesting "a party in my honor."

Americans are demanding a new way of death, and funeral directors must provide it—or die trying.

Fiddle-de-dee. It's no secret that funeral businesses loathe change, that the words *tradition* and *history* are usually engraved right on the company letterhead. I have come to this convention because I want to see how the industry is faring in these revolutionary times. I have burning questions to be answered, albeit not quite along the muckraking lines of Jessica Mitford. Mine are: Would there be spittle-spewing shouting matches between the old guard and the new? What kinds of earthshaking new products would be on show at the giant expo? Why is Greg Gumbel one of the keynote speakers?

I find my way into the belly of the convention, on a subcontinent of the Bio-Dome called Magnolia. It's teeming with 3,500 conventioneers. These sure don't look like the faces of revolution: The vast majority are white, male, and AARP-eligible. The outfit of choice is a somber suit or a white turtleneck paired with a sport jacket. A sign asks that all pagers and cell phones be turned off in the workshops. Pagers? Who carries pagers anymore? (Almost every funeral director across the country, it turns out. Old habits die hard.)

I head to the expo floor and am instantly overwhelmed. Nearly 450 merchants sprawl across 108,000 square feet of stadium-style space. There are rows and rows of booths, like a carpeted crop field tended by swarms of well-dressed field hands. Colored lights and harmonic singing erupt from what sounds like a Broadway-type performance from the casket displays in the back. A trim blonde in a T-shirt promoting a funeral-marketing company glides by on a Segway scooter.

Death merchandise is remarkably varied and, to me, an outsider, endlessly fascinating. I try to focus on the newfangled and modern but find myself mesmerized by, say, demonstrations of embalming tools. There's plenty of traditional merch, including marble urns and burial wear and hydraulic body and casket lifters. I admire the pretty display of Frigid Fluid, a leading purveyor of embalming juices, lined up in bottles of bright pink and orange; I sample Ethereal Cosmetics, whose pancake "applies smoothly and easily," just as the salesman says, "with never a cakey look or feel." I finger the racks of burial clothes—polyester dresses in shades of rose or powder blue that drape like aprons on a stiffened corpse—checking to see if they carry catchy labels like Finish in Fashion or Six Feet Underwear. (They don't.)

There's not much that's groundbreaking in the hearse universe, either, though a crowd of men gawk like boys at the gleaming fleet of silver-and-black coaches. I've heard of some out-of-the-box funeral directors using Volvo station wagons and VW vans to cart bodies, but the vehicles here are Lincoln Cadillacs or Sayers & Scovill ("the standard of excellence since 1876"). There are other familiar companies at the expo, names I didn't formerly associate with the funeral biz: Oreck vacuum cleaners are a staple of appearance-obsessed funeral homes; Hallmark hawks a line of bereavement cards called Lasting Expressions.

While the traditional merchants dominate the floor, the expo is in fact hosting 150 merchants appearing for the first time this year—almost all pushing the option to personalize. Just in time for public demand, technology has made it easy to customize just about anything—candles, tombstones, lap blankets—with a photographic likeness or quote or life story.

Cremation merch is in the greatest abundance, for obvious reasons. Even the most head-in-the-sand funeral director can recite the statistics from the Cremation Association of North America: 25 percent of Americans who died in 2003 were cremated; by 2025, it'll be 48 percent. Within a couple of decades, says CANA, *half* of all dead bodies will be burned rather than buried whole.

About 3.2 million are projected to die in the year 2025, so that makes 1.6 million cremations, or 1.6 million families looking, potentially, for creative things to do with their loved ones' ashes. And cremated remains ("cremains" in industry-speak) being the malleable, portable, scatterable substance it is, the possibilities are endless. It's enough to make an endtrepreneur out of anyone.

One might start with urns. The expo is a veritable urnotopia; after all, 56 percent of those choosing cremation will purchase an urn, according to a 2005 study by the Funeral and Memorial Information Council. There are the traditional urns made of amber- or celery-colored marble and shaped like covered vases; there are wood urns that resemble shoeboxes with intricately carved lids.

Then there's the biodegradable urn in the shape of a giant seashell that dissolves when thrown in the water. There's a paper urn packed with wildflower seeds that will bloom when buried, fed by the calcium-rich cremains. There's a handsome urn whose very chatty creator boasts it is the only one that can pass through airport metal detectors; it comes in a selection of pretty cloth wrappings, including one designed especially for the military dead. There's supposedly a guy selling urns made out of motorcycle engine cylinders, but I never do find him. (I do, later, online, at RidersLastRest.com.)

The trend toward cremations has fired up a subindustry in memory preservation. Funeral directors talk about something called a "memory picture": the last look at a loved one—embalmed to peaceful perfection and lying in the casket—that will sustain the grieving survivor forevermore. With bodies being sent straight to the crematorium, they argue, families miss out on that comforting last look.

But here's what I think. In this age of accessible technology, doesn't video provide a far more satisfying, accessible, and longer-lasting memory picture? Dozens of entrepreneurs have the same theory, apparently, judging by the booths hawking software for digital slide shows or biopics of the deceased, as well as the flat-screen LCD TVs on which to show them.

Even the flesh-and-blood memory picture is improved by technology. One company gives new meaning to the term *restorative arts*—industry euphemism for the practice of gussying up a corpse for viewing—with its cosmetic airbrushing tools. The display includes fake rubber faces that belong alongside Halloween masks—one with black patches of gangrene, another yellowed with rot around the eyes, another completely gray with oily black stains. The merchant expertly sprays pinkish pancake onto the gangrene until it disappears. "Don't let discolorations catch you by surprise!" reads the sign.

———

The conventioneers walk slowly up and down the aisles, examining the wares. For merchants, this convention provides their once-a-year opportunity to make a first impression. It's vital they do: 43 percent of funeral directors say their new ideas for products or services come from these events, according to a February 2005 survey by FuneralWire.com.

Of course, trolling the expo isn't all the undertakers do at the convention. Some meet for a morning Harley ride around town or for a brown-bag lunch to discuss scuba diving. A few seem to spend an awful lot of the time hanging out at the sports bar and watching the World Series. Many attend motivational speeches like one called "Be InVinceAble" given by Vince Poscente, whose invincibility apparently does not extend to spelling. And most sit in on at least one of the dozens of workshops, if only to help maintain their licenses by scoring some continuing-ed points.

The workshops cover everything from sales ("Keys to Customer Loyalty") to workplace behavior ("What Exactly Is Sexual Harassment?") to the funeral director's own psychological health ("Compassion Fatigue: When Caring Gets to Be Routine").

In a sign of the times, the lineup is dominated by titles that betray a slightly anxious call to get with it. "Event Planning or Profound Rite of Passage? A New Society, a New Paradigm" is one. "Strategies for Surviving Until 2020" promises another. "Ethnic Funerals: Are You Ready for Diversity?"

I duck into one titled "Promoting the Value of Ceremony to Today's Customer." It's led by Bill McQueen of Anderson-McQueen Funeral Home, based in St. Petersburg, Florida. His trailblazing business model consists of setting up storefronts in malls and promising speedy, affordable service.

This kind of talk is anathema to this service-devoted industry. But McQueen gets the attention of the packed hall by citing his business's extraordinary volume (ninth out of Florida's 883 funeral homes, he says) and breathtaking profit margin (24 percent, he says).

The secret to his success seems to be his unapologetic use of clever sales tactics. For instance, his storefronts display merchandise kiosks labeled "Best," "Better," and "Good," which seems to nudge customers toward the highest-end option. He says he makes customers wait a while so they're forced to browse the kiosks, then sends in a "funeral concierge" who shows an expertly produced DVD in which mourners talk about having "one shot" to memorialize your dead. McQueen suggests ways to encourage families to "supersize" the merchandise, especially if they're "vigilante consumers" who insist on thrift.

The tactics work beautifully: Though cremation can be had for $3,048, 16 percent upgrade to $5,479; the average burial package costs $6,408, but 22 percent choose a $9,649 option.

I'm disappointed. There *is* value in ceremony, and I believe it's no sin to gently remind grieving families of that. The setup espoused here seems not to help consumers decide what they want, though, but to steer them subtly toward spending more money. To me, that's vigilante marketing.

I look around to see what people think. There's a lot of note taking.

The workshops are not all about marketing. One called "Creating the Ulti-mate Celebration of Life Experience" teaches funeral directors how to feel.

The presenters, Valerie Wages and Tom Fulton, wear clip-on micro-phones and stride from stage to audience like televangelists. Wages com-mands the performance. She delivers a dramatic thesis statement in a Texas-flavored rasp: "We have a responsibility to *change*." She tells of a re-cent funeral for an infant in which the young funeral director searched on-line for an appropriate tribute, coming up with an oft-replicated poem titled "Little Footprints." She beams it up on PowerPoint:

How very softly
you tiptoed into my world.
Almost silently,
only a moment you stayed.
But what an imprint
your footsteps have left
upon my heart.

Never mind the awkward meter and misplaced punctuation. The point is made: A little effort by the funeral home to personalize a service can go a long way.

The undertakers in the room are moved to silence. Some women dab at their moistened mascara. One sitting next to me hurriedly copies the poem into her notebook. (Wages says several times that its author is anonymous, but later when I Google the poem, I find it all over the Web attributed to a Dorothy Ferguson.)

"When words are inadequate," Wages says in a low voice, "have a ritual. We saw that with 9/11." A slide show of images from 9/11 follows. "En-gage the five senses," she instructs. "Did Mom make doilies? Lay them out—let people touch. Did she bake? Put her chocolate-chip cookie recipe in the memory folder. Lay out the pipe tobacco or the baby powder." She shows photographs of a horse-drawn hearse for the funeral of a fire chief; a cowboy's coffin laid out amid haystacks, saddles, and a ten-gallon Stetson; a balloon release in a school parking lot for a longtime teacher.

It's so much easier, Wages acknowledges, to present a cookie-cutter "greet 'n' weep" when the alternative is to actually get involved in the life of someone who's died. "Funeral directors don't like to go there emotionally with families," says Wages. "They say, 'Oh, I would never ask those questions—they're too personal.' *Too personal.*" She shakes her head.

It is just dawning on me that funeral directors have a reason for eschewing personalized services that has nothing to do with expense or hassle or tradition. Dealing with death, day in and day out, even for those with funeral service in their blood, is utterly emotionally depleting. Following the routine is hard enough. Personalized service, by definition, requires undertakers to get involved with the deceased, to learn his or her story, to delve into the survivors' sadness. It could just break you.

Wages makes her case. "But, ladies and gentlemen, that's what funeral service *is*," she says, her voice catching. "If we don't get back to that, we're not going to be around. They're not going to want us. They're going to go off and do it themselves. Dangit, we need to get out of the box and *save* families. *Take* that step. *Make* that difference. Bring their pet in and let it sit on her chest. Who's gonna remember the *casket* she was buried in?"

Now there are more than a few handkerchiefs flapping about. She asks the audience to write a note to a lost loved one on lavender cards, seal them with heart-shaped stickers that read, "I had a personal experience with Val and Tom," and drop them into a wicker basket by the door. "You Raise Me Up" swells over the sound system.

"I *love* what we do for families, don't you?" says Wages. "What would they do without us?"

———

There are twenty-two thousand funeral homes in the United States, according to the NFDA, 89 percent of them family-owned, many for generations. It's not easy to be the one to break tradition, to go against the grain. Leading the charge are a handful of well-known businesses. Panciera Memorial Home in Hollywood, Florida, designs elaborate sets to stage theme funerals. Bradshaw Celebration of Life in Stillwater, Minnesota, is a sunny, high-tech, $5.5 million facility equally suitable for funerals as for weddings. A new facility owned by Carmon Funeral Home in Avon, Connecticut, has hired a full-time event planner.

Ippolito-Stellato Funeral Home in Lyndhurst, New Jersey, would not by any stretch claim to be on the cutting edge. But its owner, Lou Stellato, is the first to admit that the limited options of his father's era just won't cut it in today's world. He's a funeral director eager to embrace the future—with customers who prefer the past.

I had been trying for some time to locate a forward-leaning funeral director in my area of northeastern New Jersey. This turned out to be a tough task; it's a region with pockets of old-world immigrants, and funeral homes are surviving just fine by offering tradition, thank you. As for personalization, one third-generation owner told me, "Most people just don't want to bother."

From my first phone conversation with Stellato, though, he makes clear he is ready to leap, come what may, into the future.

"Oh, funeral service as we know it is over," he declares. (He is, I will learn, habitually declamatory.) "We can no longer deliver funerals out of a cookie cutter. We must become event planners. Everything has changed, and we must, too."

This is remarkable language coming from a son of Lyndhurst, the proudly blue-collar home to generations of Italian and Polish immigrants. Here, old ladies wear veils to mass. Streets crawl with Cadillacs. Spaghetti and sauce is called macaroni and gravy. Lyndhurst doesn't like change. Stellato should know: He was mayor for fifteen years.

Lou Stellato is a thick-chested man with perfect hair and rimless glasses. He speaks rapidly and confidently in the manner of a practiced politician. His manner is professional and brisk, yet solicitous. In the many times we met, I never saw him dressed in anything other than a spotless suit.

Stellato is born for this work, though he wasn't born into it; his father was a laborer for General Motors. After two years in the air force, Stellato studied mortuary science at the University of Minnesota. His degrees hang on the wall of his basement office in Lyndhurst, above a collection of Paul Anka and doo-wop CDs. One wall is covered with posters from *The Godfather*, another with prints depicting an embalming in ancient Egypt.

If funeral service wasn't always the Stellato family business, it certainly is now. When we meet, his son, Lou Jr., is midway through mortuary school; his daughter Dorianne, a nurse, is planning on starting her degree soon. His other daughter, Tracy, is an assistant principal but comes home on weekends to lend a hand. His wife, Linda, is the resident restorative artist. All except Tracy live up a carpeted stairway on the upper floor of Ippolito-Stellato, an imposing two-story building on a main thoroughfare of Lyndhurst.

The street once housed a thriving row of death-related businesses. But as funeral needs changed, so did their fortunes. Change was good for upscale restaurants like Michael's and Vivo's; Stellato sets up postservice repasts there for families who no longer want to host such gatherings in their homes.

Change wasn't good for the florists, no fewer than eight of whose storefronts once lined the stretch before Holy Cross Cemetery, according to Stellato. It's down to two. "In lieu of flowers" is to blame. The phrase has grown so popular in obituaries that it has decimated the floriculture industry, so much so that its lobbyists reportedly pressured newspapers to strike the statement from all death notices.

Flowers, in fact, have become almost passé at funerals. About.com's "Death and Dying" site featured an article in April 2005 titled "Seven Alternatives to Sending Flowers for a Funeral." It begins with a derisive comment: "Everyone sends flowers to a funeral. Sometimes families can become deluged with plants and flowers of every imaginable size, shape and arrangement. Graves sometimes overflow with sprays and wreaths." The article says some new immigrants like the Mien—an ethnic group forced from Laos, of whom about forty thousand have relocated to the United States—consider flowers at funerals offensive.

There are no Mien in Lyndhurst, as far as Dennis McSweeney knows. McSweeney has made floral arrangements at Lyndhurst Flowers since, "oh, before I was his age," he says, jabbing a thumb at his son, Dennis Jr., who is jawing with his high school classmate, Lou Stellato Jr. McSweeney is stabbing palm fronds into green foam blocks for sprays that will go atop a casket at a Stellato funeral later that day.

When he finishes, McSweeney joins me in the front room, where the shop displays its most popular funeral arrangements. We admire the white carnation cross ($200–$500) and the yellow mum bible ($250) and the red rose bleeding heart ($300–$500). They're propped on spindly metal easels that can be collapsed for easy removal from viewing room to cemetery burial site, where they're typically abandoned until the cemetery staff tosses them in the trash heap.

Arrangements for funerals used to make up two-thirds of the business, says McSweeney. Now it's half, at best. With viewing times slashed or even eradicated in the case of cremations, flowers play a far lesser role in the funeral service. Folks today request less expensive vased flowers or plants the family can take home. Some even order arrangements online.

Fran O'Rourke is in his police officer's uniform, tapping away at a computer in the arrangement room. O'Rourke owns Lyndhurst Flowers. He inherited the fifty-six-year-old business from his grandparents and has worked there full-time since his high school graduation; there's a photo of him as a kid clambering on McSweeney's back. But eight years ago, "I read the handwriting on the wall," says O'Rourke, forty-one. He joined the police force for the steady pay and the benefits. "I got a family. I gotta think of them."

<hr />

Change was shifting the foundations of the funeral business in Lyndhurst. Then two buildings on the horizon came tumbling down, and things changed instantly, forever.

On that clear day in September 2001, when two miles away across the Hudson River those shining towers fell, the American public's idea of life and death crashed and burned, too. For days, for weeks, for months, the local newspapers told of death—of final e-mails and heroic sacrifice and body counts.

The horror rippled across the country and the world, but here in New

Jersey, in the very shadow of the towers, death felt so close, so stunningly close. Eleven lost from Glen Rock. Twelve from Ridgewood. From Lynd-hurst, three.*

The funerals began even while the missing-person flyers still plastered Union Square and St. Vincent's Hospital. Stellato ran or was involved in about a dozen services, he says. But the business was far from routine. For one thing, most of the deceased were young, "in the prime of life," says Stellato; for another, the circumstances of their death made them—"for want of a better word"—celebrities. Those two facts meant any service would draw unprecedented crowds.

The more pressing question for many funeral directors was this: How do you memorialize and bury the absent dead?

Many of the victims were never found whole or never found at all but for a shard of bone or a tooth or a scrap of skin, identifiable only by DNA analysis. Under 300 bodies were found in one piece, according to newspaper accounts; of those, 12 were identifiable by sight. Of the rest, authorities recovered 19,893 individual body parts, 200 of them belonging to one man.†

Those families were lucky. Of the 2,800 who died, no trace was ever found of about 45 percent. That's 1,268 families with nothing at all of their loved one left. One couple profiled in newspapers, having waited two years, resorted to burying a vial of blood their son had donated.‡

How do you memorialize a loved one who died this way? How does routine even begin to suffice?

The answer, for some families, was to hire party planners. It was the first time event planning and funeral directing overlapped in the memorials of ordinary people, say professionals in both industries, with one expected to do the work of the other. Event planners rented giant tents to accommodate the one thousand people who came to commemorate an investment banker, according to Stellato. They organized catering and music and special tributes. But while they knew where to rent chairs and how to hire a

---

* "Tears and Tributes: Loved Ones Lost, Traditions Born," *The Record* (Bergen County, N.J.), September 2004.
† "Closure from 9/11 Elusive for Many," *USA Today*, September 2003.
‡ Ibid.

band, they didn't necessarily know where to find a registry book that wasn't frilly and wedding white. They didn't know how to order prayer cards. For that, Stellato says, they called him.

That, he theorizes, was the beginning of a collaboration that would change death service forever.

"The funeral planner," says Stellato, "was born."

Richard Aaron is an event planner. It would not be hyperbolic to call him an event planner extraordinaire. Aaron has planned and pulled off thousands of glittery events for the likes of Donald Trump and Michael Dell and the queen of Spain. On a chilly Tuesday in May, in an elegant (if not glittery) hotel meeting room in Avon, Connecticut, he is introducing funeral directors to polka dots. Displaying a slide photo of a banquet room bedecked in black circles, he pauses dramatically.

"Polka dots," he says, "equal *energy.*"

Richard Aaron, event planner extraordinaire, is teaching funeral directors a few tricks of the trade—to become, if you will, funeral planners. Undertakers from around the country have traveled here for a weekend seminar called "Exceptional Funerals: Personalization Beyond the Basics." This is not Aaron's usual gig. He is the president of BiZBash, a Manhattan-based publisher of a newspaper for the event-planning industry. Last month, he presided over an industry award show held at a blues club in Times Square, where waiters dressed in drag circulated with canapés as he presented such honors as Best Gift Bag.

"When the National Funeral Directors Association first called me to conduct a seminar, I thought it was a joke," he confides.

Eight appearances later, he's no longer laughing. He now embraces the connection between parties and passings. "All we do is party. We celebrate birth. The bar mitzvah. Sweet sixteen. Weddings. So then we die. We've had a celebration culture throughout our lives. Why not the final celebration?"

Aaron looks prosperous and buttoned up, easily mistakable for one of the undertakers if not for his exuberance. A former Broadway actor with the energy of, say, a polka dot, Aaron waits by the doors while he is introduced and then rushes into the room like a high school football star. When he reaches the podium, he whips out a wand that, when shaken, ejaculates a shower of confetti. I'm not the only one who's taken aback. The funeral directors pick bits of colored paper out of their coffee cups and look confused.

"Janitors hate me," Aaron says brightly.

He holds up a newspaper article about a Halloween-themed wedding. COUPLE SAYS "I BOO," reads the daffy headline. He displays another about an adventure race in Florida that involves some sport called bushwacking. Then another about Disney's booming wedding business. "You're thinking, What the heck does that have to do with me?"

We nod.

"Well, these people are going to *die* someday!" He looks triumphant. "These are your clients. Times have changed. Your business is changing—if you let it."

With that, Aaron launches into Event Planning 101, letting loose with rapid-fire industry jargon. He talks about the tempo and pacing of an event, the "hot spots" in a room, scripting, staging, and soundscaping. At first, the funeral directors look flattened as they flip to a glossary of terms included in the seminar packet, silently mouthing "gaffer's tape" and "mirror ball." A marquee, according to the manual, is a long, narrow tent that shelters walkways; a masking is a "scenic, black velour drape used to obscure undesirable views from the audience." Who knew?

Aaron tells the funeral directors to keep prop closets with ready-made sets to fit certain themes, like the casino funerals offered by one Las Vegas funeral home. He stresses the need for Porta Pottis at outdoor memorial events, throwing in a tip about a company that rents fancy ones with marble sinks. He warns of the risks of holding a service in a tent when it rains: "God forbid it drips on the casket."

As an exercise, Aaron asks the funeral directors to design a funeral for a celebrity. One group chooses Martha Stewart, whose send-off would feature elaborate catering, decorations from every major holiday, goody bags of seed packets, and prayer cards printed instead with favorite recipes. They refrain from harping on her criminal past.

My group is not so decorous. After much snickering about gloved pallbearers and white Bronco processionals, we propose a rather tasteful event for O. J. Simpson, with four stations highlighting his roles as football icon, actor and broadcaster, golf enthusiast, and father.

Aaron applauds us. Jim O'Boyle of Mystic, Connecticut, shrugs. "We took a stab at it," he says.

Aaron concludes by proclaiming the funeral business has taught him a thing or two about event planning: After seeing acrobats dangling over caskets, Cirque du Soleil–style, at a funeral directors' convention, he borrowed the idea for BiZBash's awards show. "I'm not saying death is theater," he says. "But it could be."

⋯

After the seminar, the funeral directors congregate for Southern Comfort Manhattans in the inn's dimly lighted restaurant. Most have just met, but funeral directors are in general an amiable sort and easy company amid their peers. Most, to my surprise, represent decades of tradition. There's Phil Conway, third generation, from Peabody, Massachusetts. There's Bernadette Officer, third generation, St. Louis. Someone in the seminar had said they were *fifth* generation.

They came to learn new tactics, new tricks to shake up staid family businesses, and many buy Aaron's message kit and kaboodle. "We have to accept that we're event planners now," says Tim Copeland of New Paltz, New York, second generation, during an earlier break in the seminar. Already he's held Harley-Davidson–theme funerals and recently begun stationing masseuses in the waiting room outside a wake. "Baby boomers come in knowing what they want, and we have to know how to offer it."

The cocktails set in, and the group begins to unwind. Undertakers tell great stories. They've always had excellent material, of course. But now their attempts to go modern are presenting challenges and creating bloopers they never anticipated.

Families, for one thing, are not what they used to be. Funerals bring together every branch of the dead person's life, which today could include multiple sets of ex-spouses and children, along with their attendant neuroses and baggage and complicated requests.

Phil Conway tells the table of a midceremony smack-down between a wife and a six-foot-three-inch ex-girlfriend. Another funeral director recalls an encounter with a widow who demanded the immediate return of the deceased's penile implant. "I have my reasons," she said. (That funeral director asks not to be identified, for fear of offending the offensive family.)

Bernadette Officer's business serves an African American clientele, among whom funeral tradition is deeply rooted and propriety highly prized. Still, with so many families in her community being broken or nontraditional, she thought nothing when a well-dressed young woman arrived to make arrangements for her long estranged mother. Then the woman laughed. "It was real deep," says Officer. The sound so took her by surprise that she looked up from the paperwork and noticed the woman's neck—"all bumpy-like and rough." She looked down at the paperwork under survivor's name: Desmond. Officer threatened any staffer who dared snigger, and the transvestite braved the funeral with a chihuahua on her arm and her head held high.

"We can put the fun in funeral," as Sam Carlisle of Mooresville, Indiana, says to me earlier. "But we can't put the fun in dysfunction."

Even for these open-minded funeral directors, adjusting to the new demands of modern funerals is a challenge. It requires skills they never learned in mortuary school and contacts they never acquired at their father's elbow.

It's one thing to agree to such unusual touches as butterfly releases and Humvee processions. It's another to pull them off.

The idea of symbolizing the deceased's ascension to heaven by hoisting objects or animals into the air has so taken hold of the public's imagination that it has become one of the most popular new features of funerals. This trend bedevils funeral directors, who on the whole are more used to placing things in the ground.

Take balloons. Ken Pescatello of Carmon Funeral Home in Avon, Connecticut, once handled a request to somehow involve a hot-air balloon in the service. Pescatello, an aggressive proponent of the personalized ceremony, dreamed up the ambitious and thoroughly impractical notion of using the balloon in place of a hearse—in other words, flying the casket in the hot-air balloon from funeral home to cemetery. The lawyers' alarm brought him back to earth, and he settled for the no-less-spectacular effect of sending an empty casket aloft to symbolize the journey.

Balloons of the normal-size variety can pose headaches for the hapless funeral director as well. Mylar, out of which many balloons are made, fails to decompose—and, when ingested, kills birds. Releasing Mylar balloons into the air may run a funeral director afoul of the law in some states.

Balloons are nice. But when it comes to symbolism, there's something attractive about incorporating a living thing into a ceremony for the dead. Butterflies, for that reason, have grown as wildly popular a feature at funerals as they are at weddings, evidenced by the dozens of Web sites offering them for this purpose.

But funeral directors inexperienced in the handling of delicate winged critters may not consider small details like their adaptability to, say, a New England cemetery in December. More than once, this oversight has resulted in the graveside release of a box full of dead bugs.

No symbol strikes the mourner with more power than the image of a white dove accompanying the loved one's soul as it flutters toward heaven. Like butterflies, doves—actually, white homing pigeons—are hugely popular at weddings as well and are offered by hundreds of companies nationwide for $100 to $500 per release. Unlike butterflies, most don't die on contact with cold weather.

But pigeons smell. And poop. And worse.

Officer, who nurses a mild bird phobia, tells of the case of the tragic dove release. A flock was brought by cage to the burial site, where, after a ceremony, they were released into the air by their trainer. Out of nowhere, a hawk swooped down, snatched a pigeon, and torpedoed it to the ground. There it began its lunch as the horrified mourners watched.

I finish my drink and say good night. These funeral directors face a brave new world. But I know that to reach its shores, I will have to move beyond their company. Change lies not within the funeral industry, but with outsiders, merchants who seek to make over funeral service itself.

Enter the end-trepreneurs.

Biodegradable You

"GREEN" BURIALS, THE NEXT BIG THING IN CEMETERY TRENDS

*Here lies Mike Megel, Shirley's husband, his only mark on earth a rock
and some ferns near a creek in Westminster, South Carolina.*

*I*t's a fine day in May, and Dr. Billy Campbell, his wife, Kimberley,
and I are trudging down a path of newly laid mulch into a sun-
dappled forest. The trail slopes gently toward the noisy babble of Ramsey
Creek. The doctor points out yellow passionflower and pussytoes. At the
creek, we rest on a stone-slab bench under a pink cloud of mountain laurel.
We watch the rippling water. Monarch butterflies waft by; a fence lizard
slips underfoot.

It's the kind of place where you could spend eternity.

I am in South Carolina visiting Ramsey Creek Preserve, the country's first "green" cemetery. In simple terms, that's where bodies are buried in their natural state: no embalming, no upright tombstones, no fancy caskets—in fact, often no box at all but a shroud. They dig a hole, put you in it, and soon you're food for wildflowers. Ashes to ashes, dust to dust. What's more, your interment there helps conserve the ground above and around you.

Never heard of a green burial? Ah, but you will. With its combination of environmentalism, whimsicality, and affordability, it strikes me as the next big thing in the death industry. When the AARP surveyed members following an article on green burials it ran in its bulletin in 2004, it found a stunning 70 percent desired such a disposition for themselves. About 19 percent wanted cremation instead, and 2 percent wished for something exotic like having their ashes shot into space. A mere 8 percent said they would settle for a traditional burial.

If green burials are a burgeoning movement, Billy and Kimberley Campbell are its placard-waving pioneers. Here's what I knew about them before my visit. Billy was born and bred here in the tiny farming town of Westminster, South Carolina, population three thousand, and serves as the town's only doctor. Kimberley, I learned when she finally returned my phone messages, has a British accent. They are gung ho environmentalists who founded Ramsey Creek Preserve as an out-of-the-box way to conserve land. I picture hippie geezers, probably curmudgeonly, the kind of interview on which I should not chew gum.

I am not looking forward to it. The day begins badly. I had extended to my husband's Aunt Rita the attractive offer of driving us to Westminster from Atlanta so she could baby-sit while I met with the hippie geezers, who, being curmudgeonly, surely wouldn't take to a reporter with a baby on her back. Rita, who is one of the world's great people, had accepted. We stop at the veterinarian's to drop off her cat, Tuna Patches, for boarding.

"She's breathing funny," says Rita, frowning.

No, she's not, I think. Rita is an AIDS nurse and former Georgia state legislator with a past in union organizing and communism. Her voice sounds like Marge's sisters on *The Simpsons*. In other words, she's not a woman given to hysteria. Still, her animals are her children, and she obsesses over their health—especially that of Tuna, who has had heart problems. She scoops the cat out of her cage and speeds into the vet's office. I sigh. We were going to be late.

Within minutes, Tuna is dead.

My first reaction, if I am to be honest, is annoyance. Yes, I am that small. Death is a disruptive event; it interrupts planned road trips and imperils baby-sitting. Never mind that Tuna's death is disrupting Tuna's life most of all and, more important to me, Rita's. I am not thinking that way. My next reaction, if I am to be honest, is amusement. A death on my way to cover death—you can't make this stuff up. I am not, as you might guess, a cat person.

And then I see her. Tuna is lying on her side on the metal table in the examining room, her legs stretched out before her. The sight of her body, so recently stilled, is inexplicably and unarguably sad. Rita is crying and stroking the cat's gray and white fur. Through the fog of shock and grief, Rita breathes shakily and asks, "What do I do now?"

It is not a rhetorical question. I have not yet had the experience of being the next of kin, but when I am, I imagine that one question will beam like a lighthouse through the despair and focus my wandering mind. We need to know: What do we do with the body?

It matters, somehow, what we do with the body. Perhaps it matters more to the survivors, but plenty of us care about our own disposition (as it's called in funeral lingo—not disposal, which means the same thing). This is one reason millions of Americans, according to the Federal Trade Commission, have signed contracts prearranging and sometimes even prepaying their burial and funeral arrangements: We want to control if not the when, then the where and the who and the how.

Our burial choices are often not choices at all, but dictated by religion and custom. Whole-body burial for Catholics; a scattering of ashes on the Ganges for Hindus. American custom has for nearly a century involved a casket placed in a hole, and it still does for three-quarters of our dead. Death industry experts predict cremation will draw even with casket burial in as little as twenty years. Yet a small but growing number seek another alternative. These are the people who really, really care not just how their bodies are laid to rest, but how its disposition might better or at least not harm the earth. These are the people who make the pilgrimage to Westminster.

I deposit my red-eyed aunt and my baby daughter at a scruffy Days Inn and drive over hilly roads through the small but not ugly town of Westminster. I pass a Confederate flag, round a bend at the Krazy Horse farm ("We Rent and Sell," reads the sign), turn right at the Church of God Calvary, and come upon a driveway with a pretty woman sitting in the gravel.

Her spiky dark hair reminds me of Pat Benatar's, and she is wearing fashionably oversize sunglasses and a Pucci-print tank top. Kimberley Campbell leaps up to embrace me. Her husband, Billy, hangs back, leaning against a white BMW convertible with its top down. He's wearing jeans and a worn cap that says "Life Is Good." He takes off his rimless spectacles and wipes them on his Society for Ecological Preservation T-shirt. When he finally approaches, I spot a sliver of a gold hoop in his left earlobe. She is forty-six; he, forty-nine. The Campbells may be hippies at heart, but the geezerly curmudgeon stereotype I can lay to rest.

We climb over a hill past a row of rhododendron, and the thirty-eight-acre property spreads out before us in its Merchant-Ivory glory. Next to an English garden crawling with roses is a house where the groundskeepers—who are also Kimberley's parents—live. Nearby is a tumbledown barn. We are walking across a clearing, and Billy is telling me how the barn will be renovated as a visitors' center with computer learning centers, when suddenly we stop.

I look down and hop back a little: I am on top of someone's grave. Albert (1923–2000) and Betty Keogler (1924–2002) are buried here at my feet, their resting place marked only by a flat, rough stone engraved with their names and dates. Had Kimberley not pointed it out, my right heel would have landed right on it.

Fifty people were buried at Ramsey Creek Preserve when I visited, with another fifty on the waiting list. Most plots are marked as the Keoglers' is, by only an unpolished rock and a bevy of carefully selected wildflowers. Some rocks have the names and dates of the deceased engraved on them; others are blank (I learn later this has to do with the lackadaisical pace of the local engraver). Some plots aren't marked at all. By the end of our hike, I learn to recognize the graves by their slight mounds of dirt and lively ecosystems, but the first few I pass right by.

Throughout the hike, which is strenuous at times, Kimberley keeps up a cheerful patter. She hops along nimbly even in her weekend-on-the-Riviera getup; she and Billy hike the trail every day, checking up on the residents. Billy strides ahead, calling out the proper names of flowers—*Passiflora lydia, Echinacea laevigata,* coreopsis yellow. "There's the plantus-I-don't-know-name-us," adds Kimberley. Billy plucks a handful of sticky-ended stems called monkey spears and aims them at his wife. She is the charmer of the two. The comforting chatter is calculated. "Cemeteries, funerals—it's like being in a foreign country to most people," Kimberley says. "There's a sense of vulnerability." She's quiet a moment. "It all sucks, basically, when some-one dies."

A few steps later, Kimberley calls back to me. "Here are the top three questions people ask about Ramsey Creek. Number one: Am I going to be dug up by wild animals? Number two: How toxic will my body be to the en-vironment—you know, will my body harm the water supply? Number three: How can I be sure that thirty years from now, this place will still be here?"

The Campbells answer these FAQs at length on their Web site, but in short:

1. Bears and coyotes occasionally roam through Ramsey Creek, but none so far have exhibited the industriousness necessary to exhume a human snack.

2. Far from soiling the soil, unembalmed human bodies make for rich fertilizer. ("Echinacea grows particularly well" over graves, says Billy; "it likes the calcium.") The creek has an *E. coli* problem, not from the bodies—the closest of which is buried two hundred feet away from the riverbank—but from deer poop.

3. Here's the beauty part. According to the Campbells' business plan, a percentage of the money for the burials goes into a trust for the preservation of the forest. Its designation as a cemetery forbids developers from building on it by state law. And the Campbells are working on getting a conservation group to hold an easement on the land, which would entitle the group to some control over development, in essence adding another layer of guardianship. The Campbells may or

may not be here in thirty years, but the trust, law, and easement help ensure that the property will.

In a circular clearing bordered by graves sits an old clapboard church. The Campbells moved the abandoned structure recently on a flatbed truck from up the road a stretch. The church was plan B; original plans for the clearing called for an elliptical stone sculpture. In retrospect, the church is a far better fit. Its presence on the property has gone a long way toward assuaging suspicious locals. Kimberley shudders. "Can you imagine what they would have said about us if we'd put up Stonehenge here?"

The church's restoration is a long-term and loving project. The antique-glass windows are made by an artisan in Texas, the carved wooden doors from a missionary church in Goya, India. The bell was donated by the family of someone buried here. The painted sign above the doors reads, "Cross-roads Chapel," with the word *Baptist* scratched out. It does resemble the chapel on Cumberland Island where JFK Jr. got married, as Kimberley mentions more than once. Once restored, the space is meant to be used for any community event—celebrity weddings, Boy Scout meetings, funerals.

In the woods, we catch up with Billy by a freshly dug mound blanketed by pine needles. Billy picks at the Christmas ferns and the ginseng and the sprig of mountain laurel growing from the swell of earth. Yvette Deschaine, the mother of three young boys, was buried here earlier this week. Cancer, says Kimberley. In the chapel, I had seen a small green album that the family must have left behind, filled with snapshots of an unsmiling woman with thick brown hair mobbed by rowdy-looking boys. I wonder how much pain it would take to keep a mother from smiling with her sons. During the burial ceremony, Kimberley slipped away with the boys to the creek for a moment of levity. The boys shouted and threw pebbles and picked mountain laurel. When they returned to the grave, the youngest, who was seven, planted the sprig on his mother's newly filled grave. "I'm sorry," he said, sobbing quietly. "I'm sorry."

We stop at another fresh grave. To the Campbells' surprise, the cemetery is attracting a growing number of fundamentalist Christians, who cite Genesis 3:19: "Dust thou art and to dust thou shalt return." "People think it's a bunch of granola-crunching hippies doing pagan things," muses Kimberley, "but most people buried here are Southern Baptists and Jews." Anna Palmer, eighty-three, was a proper churchgoing lady, the kind who dressed to the nines even to her own burial. Everyone called her Pansy. Her daughter thought she looked so fine in her blue tweed suit, white hat, and white gloves that she left the cardboard casket uncovered as they lowered her into the ground. Though Pansy and her family embraced the concept of returning her to the earth, the preacher seemed to have reservations. Looking around at the gathered crowd, he declared, "I *know* I ain't descended from no monkey." Kimberley says she gulped. Later, though, as he helped shovel dirt into the grave, the preacher recalled his grandmother's burial in a glass-covered coffin, which the funeral director had said would preserve her body for fifty years. "Even then, I remember thinking, Why?" he said.

Kimberley clucks as we approach the grave of Christopher Nichols (1975–2004). Today is the first anniversary of his death, and someone has left an expensive-looking crystal rock atop his mound, its sparkling purple jags protruding incongruously from among the ferns. Cemetery rules forbid the placement of non-native elements on graves, be they stuffed animals or geraniums. At an ordinary cemetery, the intent of such a rule would be to maintain a uniform appearance, and while appearance is not discounted at Ramsey Creek—iconography is discouraged on grave markers, for instance, a rule the Nichols family ignored by inscribing an angel and a dragonfly and the epitaph "Love Your Mother Earth"—the main point is not to disturb the carefully planned ecosystem that is the grave. "The crystal would route the rainfall away from this fern, see," Kimberley says.

Usually, alien offerings like the purple crystal are forwarded to the family. Kimberley reaches for it, but Billy stops her. "It's his day," he says softly. "Let's leave it for now."

The hiking trail has looped back to the circular clearing with the church. It is at this point that the Campbells normally talk business. Prospective customers receive the same guided hike I did, with Kimberley doing much of the chatting and Billy plucking monkey spears up ahead. Since 1996, the Campbells have fielded thousands of inquiries and several hundred visits. If you go, they will take you on the North Ridge Trail down to Mountain Laurel Alcove, where you will pause for breath on the stone-slab bench that Billy hauled here on a wheelbarrow. They will point out the uvularia and

Saint-John's-wort and potential burial plots. ("Now, that would be a nice spot," Kimberley says to me on a particularly lovely stretch of trail.)

At the end of the hike, they will tell you burials cost $1,950, with up to $325 more in fees to open and close the grave and for the grave marker. Your face may fall at the price, which, while cheaper than the $3,000 national average, may amount to more than you expected for such fundamentally basic services. "It seems hideous you have to pay to be buried, on top of medical costs," Kimberley will apologize. Then the Campbells will remind you that all profits after operating expenses go to expanding and restoring the preserve. You may take out a checkbook then and there, at which point Kimberley will push it away, telling you to think it over, to be absolutely sure.

Perhaps this explains why in nearly ten years the Campbells have buried only fifty people, while newer green burial sites around the country are quickly building their waiting lists. Sales is not their strong suit. But they're learning. The first burial at Ramsey Creek occurred in 1998; in mid-2005, the rate is two a month.

If you ask, the Campbells will explain what will happen once you are dead and ready for burial. First, Billy will photograph your chosen burial site. This is to document the wild plants so that he may return them to their proper spots after the hole is filled. Then he picks up his Truper shovel and begins to dig. Graves at Ramsey Creek are only three feet deep, not because the digging is done by a single skinny man, but, the skinny man claims, because "the point is to nourish the earth, and you don't want to sequester the nutrients." Billy says this matter-of-factly, as if a human body were only so much Miracle-Gro. Sometimes he drags a friend down to help him, but mostly it is just the doctor and his shovel.

By the time he strikes dirt, your body would have been shipped to nearby Sandifer Funeral Home. If you live elsewhere in the country—about half of those buried here are from out of state, as far away as Boston and Chicago—Sandifer would have arranged to fly your body here for $780 (your own local funeral home might have charged $2.75 a mile; from New York City, according to my calculations, that would run you $2,200). Most states demand that bodies be buried within twenty-four to forty-eight hours after death, unless they are embalmed or refrigerated. Because embalmed bodies are not allowed at Ramsey Creek, the funeral home or your hospital morgue must keep you on ice until you're lowered into the ground. (All this is moot if you choose cremation; the Campbells believe whole-body burial is far better for the environment, but they will bury cremains on the property for $500 or scatter them for $250.)

Perhaps your family will dress you in your favorite Levi's and Grateful Dead T-shirt. Perhaps you will wear your Armani suit. Perhaps you have built a casket from plans Billy has sent you, or perhaps you paid $375 to a Westminster cabinetmaker for a plain pine box. Perhaps you will forgo the casket and opt to be wrapped in your grandma's quilt. Your family and friends will attend a service in the JFK Jr. chapel or simply gather round the grave site.

Billy will use a golf cart to truck your body down the hiking trail and three cloth straps to lower your casket into the rather shallow hole. (If you have chosen to forsake the box, he will use a muslin sling devised by Kimberley.) Your religious representative will pray, your Beatles anthem will play, your family will roll up their sleeves and help tuck you into the earth. In the end, it will once again be the doctor and his shovel.

<hr />

Right now, the doctor is shoveling sautéed squid into his mouth. Billy, Kimberley, and I are at Mum's, an upscale (if chain-style) Italian restaurant. Kimberley is wearing a diaphanous top and translucent powder on her cheeks; Billy has changed his T-shirt. Upstate South Carolina is the "middle of bloody nowhere," as Kimberley puts it, apparently true as far as restaurants go; the supply does not match the demand, judging by the wait. A light drizzle starts, but we agree to grab an empty table outside anyway.

I don't know if it's the tropical night or the beer, but suddenly the laconic doctor is talking a blue streak.

Before my trip, various people who knew him had described Dr. Billy Campbell to me as "intense." Frankly, I had found it curious, if not outright weird, that a busy country doctor would take a side job as a cemeterian. It turns out he is much busier than I knew; until 2000, he was on the executive committee at Oconee Memorial Hospital as well as chief of the emergency room, all the while running a private solo practice that today counts four thousand active patients. The Campbells are raising a sixteen-year-old daughter, Raven, plus five Jack Russells and a mutt. Raven has just been accepted to an arts boarding school, and one of the Jack Russels has just had three puppies. Billy also mentors medical residents. He used to row a raft as a river guide. You *need* intensity to live that hard.

I ask why he would bother with cemeteries. He answers, "It's always been a dream of mine." He says this with a straight face, as if every little boy fantasizes about ways to bury bodies. A third-generation farmboy whose family grew cotton and raised hogs and chickens, Billy grew up certain that someday he'd find a way to protect his beloved land. An idea took seed in grade school, when a homeroom teacher announced his intention to be buried in a burlap bag under a tree. "He was a little bit of a strange guy," says Billy. "But I was thirteen, and I just thought that was the coolest thing I'd ever heard." He studied botany at Emory University until premed courses took over. Meanwhile, a human ecology class and a death-obsessed girlfriend got him thinking about how humans interact with the land.

When the concept of a green cemetery finally gelled, Billy was so excited that he couldn't stop discussing it. "What you have today is rituals that reinforce alienation from nature," he says, his words tumbling like rapids. "For people who are alienated from nature to begin with, putting bodies in a box in a box," meaning the casket that goes into the vault that often lines a plot, "helps them to deny the natural process. We need to return to simple rituals that reintegrate nature and the human world for the funeral to be a transformative and not normative experience." The idea is to "link land conservation with ritual and with people in a very fundamental way." With the shadow of big-city sprawl fast encroaching, places like Westminster need economy-proof protection from development. "You're not going to build on top of Momma." The cemetery, he adds, would become a place to catch a butterfly and wade in the river. "A place to get your yayas out. A place to connect with nature. If what brings you there is your grandma, then there's nothing wrong with that."

Kimberley raises her wineglass and leans over. "I told him, 'You have to do it or shut up because you're scaring people every time we go out to dinner.' "

So the couple founded Memorial EcoSystems, a for-profit business dedicated to creating cemeteries that will also conserve land. ("It's for-profit, meaning we're all for it—if we ever make any," Billy quips. In early 2006, he tells me they hit the black for the first time.) They raised $200,000 from family, friends, and their personal savings, paid off all the debt to convert the thirty-eight-acre Ramsey Creek Preserve to a cemetery as South Carolina law requires, and prepared to sell hundreds of plots.

They built it. And no one came.

The rumor mill pumps diligently in Westminster, where once Dr. Campbell was said to have AIDS because he suddenly lost fifteen pounds (heartsick, he says, while courting Kimberley), but moreover because he was the only doctor in the county who would treat those patients. So when townsfolk learned he was planning a natural burial ground, well, the tongue wagging grew fierce. One patient teased Billy that he was like a veterinarian with a side job in taxidermy. Others weren't so tickled. The doc was going to put dead patients in the crick, haven't you heard? He and his foreigner wife were part of some pagan cult. Oh, yes, we heard it on the radio.

One day, Billy picked up the phone at his practice and an old lady shrilled, "We don't want to drink your dead man's soup!"

He looked at the caller ID. "Mrs. Johnson," he began.

Silence. Then, "How'd you know my name?!"

Unmasked and confused, Mrs. Johnson, a longtime patient, listened as Billy explained his mission. "Once they talk to me, they see that it's not crazy or scary," he says. Still, the harrassment continued. A local paper called Ramsey Creek Preserve a "tree hugger's heaven." It got so bad that a graphic designer friend mocked up an ad for Billy to place in response:

Billy's Buzzard Ranch
Ashes to ashes, dust to dust,
Bodies to birdshit—it's all the same to us.

"We were terribly naive," Kimberley says.

But slowly, word got out. The greenie subculture began to buzz about this new concept put forth by one of their own (the doctor founded the South Carolina Forest Watch and belongs to various other environmental groups). Even locals began to embrace the idea. The tide may have turned when a man named Charles Ramey, whom Billy fondly calls "the neighborhood curmudgeon," bought a plot at Ramsey Creek. His stated purpose: to save money. Ill from liver disease, Ramey would bellow on visits to Dr. Campbell's practice: "You might as well keep me alive. I've already paid for my plot. You and I know this is the only way you'll get more money outta me."

Oddly, it's the likes of Ramey who necessitate concerted efforts at land conservation in regions like this. Westminster knows no zoning, populated as it is with folks who insist on their right as Americans to keep junk cars up on blocks in their front yards. Billy recalls that Ramey used to say, "I love the woods, but I hate those damn environmentalists." Still, Billy insists that Ramey was green at heart. "He was just the gun-toting, Confederate-flag-flying, chaw-chewing kind."

Today, over half of those buried at Ramsey Creek are South Carolinians.

MIKE MEGEL
1936–2003

*W*hen the maroon Dodge pickup pulls up to Ramsey Creek Preserve, I approach it hesitantly. I am supposed to meet a widow here, and this is no lady's ride. The burly truck stops, and Shirley Megel drops down from the driver's seat. She is a petite woman of sixty-six dressed in a sky blue pants ensemble, her short hair curled neatly, a shy smile on her face. Sure enough, the truck was her late husband's, bought just before his death. It's a gas-chugging monster, but it was Mike's and she can't let it go just yet.

Mike Megel died of cancer. He did not like doctor's visits; few lifelong smokers do. Shirley didn't nag him. One afternoon, a stabbing pain in his abdomen brought him to his knees. As Shirley watched, he knelt there in the living room—and reached for his cigarette. He lit up, then he fell over. She didn't say a word. After that, without discussion, they went to the hospital. Later, he said, "Thank you." He was thanking her for that last undisturbed puff. She understood.

By the time Dr. Campbell diagnosed him in early 2003, the disease had spread from his lungs to his lymph nodes and liver. He died that September. Mike Megel was sixty-seven.

It's almost two years later, and Shirley Megel's grief still hangs around her like a light perfume. You don't let your high school sweetheart go so easily, not when he robs you of your fiftieth anniversary by two months. The two met as teenagers in Papillion, Nebraska. Mike landed a good job as a cable technician for AT&T. The job moved them around a lot, but they didn't mind. They worked hard and raised six children.

After thirty years, Mike took retirement. AT&T's plump pension and benefits afforded them some truly golden years. Mike was an outdoorsman. The couple moved to St. Helena Island in South Carolina and managed an inn. They kept a pontoon boat, big enough for a toilet in the cabin. Mike bought a used two-seater plane. They traveled to Canada in an RV. Finally they settled down in a condo in Westminster, with its forests for hunting and creeks for fishing and proximity to their grandkids. It would be a good place to grow old together.

On a visit back to Nebraska, they visited the Megel family plot in a Catholic cemetery. They found the gravestones sunken and damaged, despite arrangements for perpetual care. "I don't want to be here," Mike told his wife.

Dr. Campbell told Mike about Ramsey Creek during his medical visits. Mike loved hearing about it. The land is fifteen minutes from their condo in town. One day they went to visit.

"This is it." Shirley has stopped walking. We are on the hiking trail about five minutes from the clearing, in the thick of the woods. All we hear are the creek and the insects. I look to my left and there is a small mound of dirt, almost flattened to the level of the land around it. There is a rock, but it is not yet engraved. This is it, the site of Mike Megel's grave.

Mike's funeral was held here that September of 2003, under the tent of trees and to the chorus of the creek. The children and grandchildren and neighbors and church friends came. A Catholic pastor who was friendly with Mike conducted the service; it wasn't a full mass, being that it was held outside a church, but that wouldn't have mattered to him. They could see the mountains through the trees. Each of them crumbled a handful of dirt onto the cardboard casket when it was lowered into the hole. Afterward, everyone squeezed into their condo to eat. "I don't know how they all fit, but they just did," says Shirley. "It all just seemed perfectly . . . natural."

She visits him here every week. Last time she brought their Jack Russell mix, Peaches, with whom Mike had a special relationship. Peaches scampered ahead down the trail, then suddenly she stopped—right in front of Mike's grave. "It was the strangest thing," says Shirley. "It was her first time here, but she just knew."

Life isn't so golden without him. Mike's pension stopped with his death, there was no life insurance, and AT&T is now yanking the retirement health coverage. Shirley lives mostly on Social Security. But a divorced daughter has moved back in, and they keep pleasant company. Shirley spends afternoons planning and cooking their diet-conscious meals.

Shirley fans her hand over the plants on the mound. "I keep meaning to plant a wild orchid here," she says, perhaps to me, perhaps to her husband. Slowly she gets to her feet. "So long, Mike," she says lightly. "Next time I'll bring Peaches."

✧

*T*he backwoods of upstate South Carolina are nice, but to meet his goal of preserving one million acres of land over the next thirty years, Billy Campbell will have to expand. No matter how intrigued they are by the concept, not many people will go so far as to ship a corpse all the way to South Carolina. "People will choose a green product if it's not too out of the way for them," says Billy. "For this to work, we need fifty to sixty locations close to people." They are currently scouting near Atlanta, Los Angeles, San Francisco, Taos, and Santa Fe.

Swept up by Billy Campbell's fast-talking optimism, it's easy to forget the logistical looniness of what he's proposing. I find this a common trait among visionaries, this spell they're able to cast with their "out there" ideas—on me, at least. Once for an article I spent two hours on the phone with a think tank economist who explained how cars built of ultralight carbon would gain three times their current mileage, after which I came away completely convinced this was the only way to solve the energy crisis. Credulity is perhaps not a trait to which a journalist should be admitting.

Billy is equally persuasive. Nevertheless, I ask: How exactly does a country doctor expect to purchase one million acres of land in urban centers? And given that cemeteries are not generally considered desirable neighbors, how does he expect to clear the legal, bureaucratic, and zoning hurdles?

His solution: partnership. Teaming up with conservation groups like the Audubon Society or Trust for Public Land would give him financial and administrative muscle. He would design, manage, and consult; the groups would own easements on land donated or bought through private funds. Revenues from burials could pay for some land purchase as for preservation. By Billy's calculation, two hundred acres of green cemetery would yield up to $100 million, 10 percent of which by law would go into an endowment fund to protect the land. By my calculation, even at the upper-end rate of $3,000 a burial, this requires 33,333 burials, or 167 bodies per acre. Billy says he hopes to keep the number well below 100 burials an acre, in part by soliciting donations from families to conserve the land. Either way, the doctor will need a bigger shovel.

Or many shovels, and shovelers to go with them. Of course, no one could do this alone. And, as a married man, Billy ought to know that partnership isn't easy. The Campbells state repeatedly that theirs is an idea they want to see duplicated across the country, backing up their claim by offering aid free of charge to those who ask. The idea, they believe, is so powerful and its potential so world-changing that it must be set free. It's not a matter of intellectual property—after all, green burial has long been offered in the United Kingdom, home to some two hundred so-called woodland cemeteries. In the United States, a number of cemeterians, entrepreneurs, or environmentalists have had the idea independently. But anyone who decides to start up a green cemetery—two brothers in Glendale, Florida, with a 350-acre farm, for instance—winds up at the Campbells' door.

Mary Woodsen, an environmentalist and science journalist in Ithaca, New York, stumbled into the Campbells, too, in her efforts to launch a green cemetery in her state. Woodsen is a conservationist and antipolluter so devout that she peels the labels off tuna cans to recycle. ("On my own I produce maybe one bag of garbage a year," she says. "I realize I'm a little over the top.") Her admittedly obsessive environmentalism led her to consider the wastefulness of the death industry. Her research turned up some astounding numbers: Americans put 827,060 gallons of embalming fluid, 1,636,000 tons of reinforced concrete, 104,272 tons of steel, 2,700 tons of copper and bronze, and 30 million board feet of hardwoods into the ground each year. With the concrete alone, I reckon that's enough to build 13,573 American houses. Through green burials, Woodsen figures, Americans can cut that waste dramatically.

For the Campbells, sharing their concept with people like Woodsen is like preaching to the choir: They get it already. But what is becoming clear is that an idea set free is an idea co-opted and changed, particularly when an entity as powerful as the funeral services industry sniffs the pungency of its potential profits. The signs are already on the tombstones. Earlier this year, Billy and Kimberley were invited to Whittier, California, by Rose Hills Cemetery, a business so gargantuan that it reportedly pulls down $500,000 a week in the sale of pre-need plots alone. Cemetery executives showed them around the 1,400-acre memorial park near Los Angeles. Here, according to the Campbells, was the fake-rock waterfall for the Mexicans; over here was Korealand; here were the $1 million family estates for the super-rich. And here, said the execs, beaming, would be where the greenies could go. The Campbells were agog. Their daughter Raven marveled later: "It's like an Epcot Center of death."

The two men from California who came calling in 2003 seemed, in contrast, to get it. Tyler Cassity was already a name in the burial business, having transformed the ghostly, sixty-two-acre resting place of Rudolph Valentino and Cecile B. DeMille into the gleaming, innovative Hollywood Forever Cemetery. Joe Sehee was Cassity's media consultant. They toured Ramsey Creek and hunkered down together at the Campbells' manor, where Billy readily shared his vision and experience.

"We all just hit it off right away," says Sehee, who had already become convinced that the business of death was in need of a drastic makeover, and not just by buffing up celebrity tombstones. "There's so much wrong with the funeral and burial business, I don't even know where to begin. It needs to be completely disrupted on so many levels. When I met Billy, everything fell into place."

Almost immediately, the trio agreed on a partnership, say Billy and Sehee. Also that year, Cassity had bought the neglected but ideally situated Daphne Fernwood Cemetery in Marin County, a suburb north of San Francisco. Marin's residents are an ideal market: They have a median income of $71,000* and a cremation rate of at least 70 percent,† marking them not only highly educated, but also socially progressive and thus more likely to consider alternative dispositions. Its thirty-two acres were originally reserved for the employees of the Sausalito Ferry Company, which I suppose was considered a job perk back in 1894. Those graves would remain undisturbed; there was enough unused land to dedicate to green burials, Cassity and Sehee suggested, and cleaning up the property would attract hikers from the abutting Golden Gate National Park. Offering nonburial functions, and thus making the land attractive for the living, is a pillar of the green cemetery movement.

Rechristened Forever Fernwood, the cemetery sits in a misty valley just north of the Golden Gate Bridge. It's fronted by a nearly windowless cement bunker that serves as the office. As I approach, a scraggly, toothless old man lurches out of its shadows. Roy is the resident gravedigger, I learn later, but I am a little spooked.

---

* U.S. census.
† International Cemetery and Funeral Association.

Outside, the building resembles a bomb shelter; inside, a spa. Its long corridor is lined with skylights, smooth stones, and plants, the rooms are painted in sea-foam tones, and candles give off the scent of wet moss. Arrangements are made in a small front room with a round table and warm lighting. An Australian shepherd mix named Owen comes to sniff my legs.

Gary McRae manages Fernwood. McRae, a soft-spoken, compactly built Scotsman, moved to the United States in 2003 after working for ten years as a homicide detective in the United Kingdom. He spent the last of those years at Scotland Yard as a family liaison officer, a role created after 9/11 to help get families of murder victims to cooperate in solving the crimes. He moved to California when he married an American girl; he was introduced to the funeral business because she happened to be Tyler Cassity's assistant at the time. McRae is now a licensed funeral director. This is not as huge a leap as I initially imagine; after all, both jobs require delicate handling of grieving people, and both deal with death.

He smiles. "But here, most of the bodies arrive in one piece."

McRae shows me a map of the property, with sections outlined in red and yellow and blue like individual countries. The blue country is where the natural burials will take place, in states with names like Free Spirit. Next, McRae brings out a binder with photographs of wildlife. "You can have a tree planted over the body, or native grasses, or wildflowers," he says—bright yellow California poppies or purple lupines or wild irises.

Mixing sentiment and technology is a trademark of the Forever network; at the L.A. location, a studio churns out Hollywood-quality biopics of the deceased called LifeStories, a much-buzzed-about innovation in the funeral business. Fernwood is going another step. Gary fetches a handheld tablet from Hewlett-Packard about the size and look of an Etch A Sketch. He touches the screen with a stylus. Up pops the company Web site via a wireless connection to the Internet. The plan is to hand out the gizmos to visitors, who can log on to webcasts of the funeral and watch biopics of the deceased. (McRae clicks open the LifeStory of Margaret Sassoon, whose husband, Moe, talks about her career as a ballerina as a tune from *The Nutcracker* tinkles in the background.) Then the visitor can use its Global Positioning System satellite technology to guide them to the grave site—a handy tool, as many of the plots will be unmarked. It's just a prototype, but it's pretty cool.

The cement bunker at Fernwood will be a sort of shopping mall for the green burial, including a crematorium with an observation room, a showroom displaying biodegradable EcoPods in place of traditional caskets, and a ceremonial space with a waterfall outside and plenty of room for rituals that involve dancing. In December 2004, when I visited, none of this had actually materialized. Still, journalists love to write about this kind of stuff, and by then Fernwood had a thick sheaf of clippings from local papers and national magazines.

Behind the scenes, all was not peace symbols and fluty music. The partners were arguing heatedly. Billy was adamant that a green cemetery must abide by certain rules, such as forbidding embalming, planting only native plants, and, most important, taking concrete steps to preserve the land. Cassity, he says, felt those rules could be bent in the name of business. Sehee took Billy's side.

By the time I met the Campbells in May 2005, the partnership had all but dissolved. Sehee says he and Cassity had parted ways earlier that year. By fall, Sehee had formed the Center for Ethical Burials, a nonprofit dedicated to standardizing green burials by offering a Good Housekeeping–style stamp of approval on environmentally minded cemeteries, crematoriums, and funeral homes. The seal would ensure, for example, that a crematorium uses filters to keep mercury in dental fillings from spewing into the air. (That's not as outlandish a concern as it sounds; in the United Kingdom, studies have shown that crematoriums are responsible for 16 percent of mercury emissions.) Some of the language on the center's Web site is pointed: "Beware of any 'green' cemetery that acts as if it furthers a conservation purpose . . . an easement ensures that what looks 'green' today doesn't become a place that tomorrow accommodates large marble monuments and embalming . . . you may be dealing with a conventional cemetery using the environment as a marketing ploy."

Meanwhile, the Campbells' for-profit Memorial EcoSystems now posits Billy and Sehee as concessionaires of green burials, advising funeral service providers on a freelance basis. Their first big project is conserving one thousand acres of ranch land near Santa Fe by offering green burials on a portion of the property; contributions made by the pre-need purchasers of the plots would pay for the rest of it.

The dream is a big, sprawling, and still evolving one. At the close of 2004, Sehee had talked of a sort of natural death center in Santa Fe, part meditation retreat, part hospice, part funeral home, with a role for Sehee's wife, Juliette, a reflexologist who was training as a death midwife. A year later, plans for a physical building had been scratched in favor of the concessionaire idea. "Funeral directors have become obsolete," says Sehee with the same conviction and passion with which he had talked initially about the death center. "What you need now are funeral agencies."

The split with Cassity left Sehee bitter, although he is probably better equipped than most to move on. Sehee is a youthful forty-four who takes business calls on his cell phone while feeding his toddler. Sehee has already lived several lifetimes. Sehee has directed a social justice program at the University of San Francisco as a lay Jesuit minister, taking students on field trips to Nicaragua and El Salvador; he's advocated for affordable housing in the Bay Area; he's reported for NBC's *Dateline* and the *L.A. Times;* he's consulted for IBM. To blow off steam, he has performed monologues in a lounge act called Joey Cheezhee. A friend who was a former Franciscan monk and aspiring singer introduced him to Tyler, who recruited Sehee as his media and public relations consultant.

"It was PR, but what I did for Tyler went beyond that," says Sehee. In his role publicizing Cassity's Hollywood location, "I think I changed the way people related to that cemetery." He vehemently stands by his claim that he brought the green burial idea and news of Dr. Campbell's existence to Cassity. A long article profiling Cassity that ran in *The New Yorker* in September 2005 chronicled and thereby cemented their divorce. Sehee felt set up and exposed by the article, but he also gained from it; the day after it appeared, he says, Josh Slocum, head of the Funeral Consumers Alliance and a highly regarded industry critic, agreed to join the board of the Center for Ethical Burials.

Cassity did not appear during my tours of Fernwood in Marin County and Hollywood Forever in Los Angeles and did not respond to later requests for interviews. But I did speak to his old friend, Ron Hast. Hast is publisher of *Mortuary Management,* a trade magazine for funeral directors, and the weekly newsletter *Funeral Monitor.* Over forty-eight years in the business, he has owned fourteen funeral homes and is regarded as something of an industry sage.

I ask him for his take on the green burial movement.

"There is absolutely no trend here," says Hast. He is speaking not just of Marin County, where he lives, though indeed he says that in his twelve years of residence he has attended exactly two funerals that involved a casket; all the rest, he says, were cremations. Cremation, he says, is the real story, the true trend that will transform the death care business. (Hast, I should note, owns the Cremation Society of Los Angeles, which arranges cremations.)

"Green burial is a novelty," he says. "A cottage item. Nothing more." As proof, he cites a Catholic cemetery in Redding where no-frills burials have long been offered but remain unpopular—far less popular than cremation. "And it's even cheaper! But people don't want it."

Hast played a role in Cassity's purchase of Fernwood, he says, by introducing him to the land. But while he still considers himself a "good friend," Hast has parted ways with his younger compatriot on the business vision for the cemetery. "He's a romantic," says Hast of Cassity. "I told him to put in a crematorium, a nice one people would want to use. In a weekend he could have cleaned up that place and had a tremendous business. But his heart is in the romance of the forest. He's spending millions to attract pennies."

Green burial, Hast says, is by no means a new concept, even in this country—yet the idea has struggled to take off. Of Billy Campbell's Ramsey Creek Preserve, he says: "Romantic theory is what it is. All the free press you lot gave him, and how many has he buried? Not one hundred people in five years. If it's so important, so desirable, so wanted, so preferred, why would he bury only one hundred people in five years? Some cemeteries bury one hundred people in a *day.*

"You say it's beautiful? Well, I'm sure it is. But what's happening out there? It's sitting there, doing nothing. I hear he's going to double his property. What for?"

❖

Maybe Hast is right. Certainly, cremation is by far the bigger trend, and the practice of burying bodies in their natural state to help conserve land may always remain a novelty, a niche. Still, plenty of others in the business look at green burial and see a different hue of green.

I recently received a press release for the Eco Casket. "Crafted from certified sustainably harvested oak, organic cotton, and an all natural finish, the Eco Casket will appeal to families that care for the environment," it read. "Make no mistake about it, this is a top notch casket in terms of quality and esthetics, and is certain not to offend more conservative members of the family."

Easily scandalized relatives might take less well to the EcoPod, a casket made of paper and shaped like a suppository, marketed by a company in London. Kinkaraco, a San Francisco outfit, sells biodegradable shrouds with names like the Botanika RestSpa. Its slogan: "Look beautiful—in the last thing you'll ever wear."

What's more, cemeteries far and wide are likely to begin offering a green option, with refrigeration in place of embalming, a plain casket sans vault, an oak tree instead of a marble marker.

On the face of it, this doesn't seem like a bad thing. The Campbells and Sehee would shout "greenwashing," and indeed, the point of those burials is clearly not land conservation, but personalization. Baby boomers who spend their adult lives separating paper from plastic demand the option of environmentally correct burials. Anything else could look hypocritical. Besides, recycling makes us feel good about ourselves, and so, too, might an Eco Casket. Is that so wrong?

It is, if we care about more than appearance. It matters what we do with the body. When we finally admit the so-called American way of death too often involves senseless waste and willful destruction based on questionable tradition, that's when we'll toss it all out and start again. Then and only then might we embrace true tradition in our death rites, doing as our forefathers did and theirs before them, by offering our dead back to the earth, with love. Ashes to ashes, dust to dust.

I don't know when or if that will happen. But it seems to me that Kimberley was right: It all sucks, basically, when someone dies. Planting a tree over a chemical-free body—and donating the four grand you could have spent on a casket to keep a parcel of land perpetually Trump-free—could make it suck a little less.

Before we leave Westminster, we stop by the Campbells' home. It is a red-roofed, English-style manor with a driveway encircling a lush rose garden and a fleet of Mini Coopers parked under an attached carriage house. Billy Campbell and Aunt Rita are instant friends, prattling about the evils of the medical bureaucracy and their respective humanitarian causes.

As we sit on the patio and watch the Jack Russells tumble toward the orchard, Kimberley asks Rita if she would like to bury Tuna Patches at Ramsey Creek, where pet burial is welcomed for $25 to $75. Rita hugs her.

"Thank you," she says. "No, I know what I'll do."

Tuna was so named because that's what lured her to Rita, who would leave a bowl of the stinky canned fish outside for the stray. Over time, Tuna slipped from the wilderness into Rita's house and her life. Rita has decided she will take Tuna's cremated remains and plant them under her magnolia tree. It is not Ramsey Creek, but it is close enough.

"She came to me from out there, and that's where she'll return. It seems," says Rita, "only natural."

Ashes to Ashes, Dust to Diamonds

HOW TO TURN YOUR LOVED ONE INTO JEWELRY, AND WHY

Valerie Sefton sparkles on in rings set with diamonds made out of her cremated remains.

A dog is yipping inside a well-kept brownstone in Manhattan's East Village. I climb the stairs to where a woman is waiting by an open door. "Come, come, Nica," says Peggy Atkinson to her Maltese, who jumps up against my legs and then skitters inside.

Peggy greets me politely, if guardedly. She is tiny, bespectacled, and dressed in a black turtleneck and long black skirt. She embodies my image of a Broadway voice teacher, which is what she is—all except for her voice, which is raspy with cold. It is our first meeting. We have spoken over the phone, and because I have asked, she is letting me witness an intensely personal event. Peggy Atkinson's husband is coming home today, and I am here to watch her open the package.

DON ATKINSON
1940–2004

*D*on Atkinson died in January 2004. His death was shockingly sudden; at sixty-three, he was as robust and blond and theatrically handsome as ever, until one night when his heart stopped working. He collapsed on the kitchen floor while getting an antacid pill at 4:00 a.m. The medics came, but Peggy knew it was no use.

After saying good-bye to his still-warm body at the hospital, Peggy had no desire to view it again. "What was Don was gone," she says simply. She had him cremated and for a while kept his ashes in a mahogany box atop her dresser. But she knew what she would do.

Not long ago, she had been reading a women's magazine—*Vogue,* or maybe it was *Better Homes & Gardens*—and had come across one of those little blurbs in the front section, this one about how they could make diamonds out of human cremated remains. Diamonds! She had shown the article to Don, and together they had exclaimed over the marvel of it. How perfect—how them. Whoever went first, they said, laughing, this was without question what the other would do.

After Don died, Peggy fished out that article. It listed a Chicago company called LifeGem, along with its Internet address. She looked it up and then telephoned. It would take eight ounces, the representative told her, gently—eight ounces of Don to make a gem.

"I didn't do it right away," she says. "I held back for a little while." Finally, with her brother's help, Peggy opened that heavy mahogany box. She cried then. "That was the hardest part, measuring and sending it. That was the hardest. Very emotional, opening up that sealed box."

Four months later, the diamond was ready. This morning at 9:00 a.m., the FedEx man delivered a box about the size of, but much lighter than, a dictionary. It is sitting here at the edge of her massive dining table.

Don and Peggy Atkinson were a classic New York actor couple. They met on the set of the original Broadway production of *Fiddler on the Roof*, Peggy cast as Chava, the younger, idealistic daughter who marries the student Fyedka, played by Don. Peggy was twenty-five, a Brooklyn-born spitfire with Italian eyes and waist-length locks; Don, twenty-seven, was a muscular dancer from Ohio with yellow curls and a face-splitting grin. They met, fell in love, fought, broke up.

It was the apartment that brought them back together. "The Lower East Side was the Haight-Ashbury of New York in the 1960s," says Peggy—bohemian and slightly dangerous. Peggy walked up those stairs and saw an entire floor of a brownstone, an unbroken bowling alley of a space, high-ceilinged, room enough for two passionate people. She marched up to Don during the next show and handed him a note. "I've seen our apartment," it said. They met afterward at the Theater Bar on Forty-sixth Street, made up, moved in. They paid $150 a month to start. It would be home for the next thirty-five years.

This being an actor's home, photos, many of them professional, cover the mantels and bookshelves. There is a black-and-white head shot of Don, the kind actors bring on auditions with their résumés pasted on the back. He's over-the-top handsome, with a broad nose, lantern jaw, Chiclets teeth, and wide-spaced, narrow eyes, like a cartoon rendition of a prince. Here is the couple, sun-streaked and grinning, wearing yellow life jackets while swimming with the dolphins in the Bahamas. I like best the close-up of Don after a day of sailing the twenty-two-footer *Pegasey* they kept at Peggy's family house on Long Island, his beard and curls whitened by sea and salt, mischief in his smile.

This is not memory lane; these photos are fairly recent. This was their life just yesterday.

On the wall there is a poster-size black-and-white portrait of the two of them on their wedding day. Peggy wears a cowl-neck sweater pulled over her head like a medieval headdress, and Don is in an Edwardian suit with ruffles. They look as though they have wandered off the set of *The Lion in Winter*. They held their medieval-themed wedding on the spectacular estate of a school for the deaf, with giant bowls of fruit and tables of pie. "Like Camelot," says Peggy.

Peggy didn't want a funeral for Don. No one who holds a medieval wedding would want a traditional hankiefest. She and Don's best friend, Ed, "designed and put together—no, that's not the right word—we *composed* the ceremony." The program was titled *A Celebration of the Life of Don Atkinson*. Don wasn't religious, but he had liked how he felt at the Marble Collegiate Church, so they held it there. Hundreds came. The show consisted of four scenes, each depicting stages of his adult life. Scene One: The Hoofer. Scene Two: The Director. Scene Three: The Salesman. Scene Four: The Man. Students sang "Seasons of Love," from *Rent*. Ed rewrote the lyrics to Stephen Sondheim's "Old Friends," from *Merrily We Roll Along*. Friends sang "Till There Was You" and "Try to Remember" and Handel's "Come unto Him." They closed with Elvis's "Blue Suede Shoes" and "Great Balls of Fire."

How many people hugged Peggy, she doesn't know. She does know she cried and cried.

❖

*T*he apartment is quiet now but for the bleating of a nearby car alarm. Peggy leads me through the place, which, like Don's funeral, is set up like a series of scenes in a play about their exquisitely eclectic lives.

SCENE ONE: The Actor. The scene is set, quirkily but charmingly, in the small bathroom off the kitchen. Its walls are plastered all the way to the ceiling with posters and playbills from all the shows they've done. Don could carry a tune, and Peggy could follow a step—"back then, you had to do everything"—but it was his athletic hoofing that led to his work with Jerome Robbins, Agnes de Mille, and Michael Kidd, and her arresting voice to her roles in *Two Gentlemen of Verona* and *Kismet*. In summers, they codirected the College Light Opera Company in Falmouth, Massachusetts. "We were," she says, "quite a team." But Broadway dancers have a shelf life, and when the calls began to dwindle, Don simply took up something else.

SCENE TWO: The Craftsman. Don loved woodworking, honing his skill building their bookcases and fireplace, and tried but failed to turn it into a business.

SCENE THREE: The Aficionado. Don loved wine and took a job at the famous neighborhood store Astor Wines and Spirits. He brought home bottles—"ruination" to Peggy's figure—and kept a log, over the years turning himself into an expert, particularly on the wines of Bordeaux, and earned the title of senior wine consultant.

SCENE FOUR: The Sportsman. Don loved to shoot, but the thought of killing animals made him nauseous. The mounted deer's head above the fireplace was retrieved from the trash. He named it Clarence and every Christmas hung its antlers with balls.

We have come to the grand piano. Peggy's second career is as voice teacher to a steady stream of students from New York University's Tisch School of the Arts. The open score on the stand is "Someone to Watch Over Me."

Peggy has cancer. She received the diagnosis the Christmas after she lost Don. She found an aggressive doctor and is battling the disease with a round of chemo. "I figure I have nothing to lose but my hair," she says, touching her ash blond wig. She smiles. "It's been a rough year."

She pauses, and our eyes rest on the FedEx box.

"Well," says Peggy, taking a deep, rattly breath. "Shall we?"

I saw a human diamond for the first time at the National Funeral Directors Association convention in Nashville.

LifeGem has an unassuming booth in a center aisle of the 108,000-square-foot convention expo, a couple of cloth-covered foldout tables with brochures—a far cry from the Vegas-style extravaganza over at the casket section. LifeGem's logo uses a quiet typeface for the company name, the dot over the "i" in the shape of a diamond with a crown of sparks over it. It's elegant and simple. In diamonds as in cremation, too much flash is ostentatious and disrespectful.

But the glass-enclosed case on a pedestal catches my eye. I park Mika's stroller to peer inside the case. There are five gems, one blue, the rest yellow. Some are cut in a round shape, some in an oval. The largest looks to be at least a carat.

A smiling man approaches. Greg Herro, LifeGem's CEO, is its public face and top salesman. He's the one they send to Japan and Australia and to talk to the press. Herro is one of four friends—two sets of brothers—from the Chicago area who have risked everything they own to launch a business selling diamonds made from cremains.

It all began with Rusty Vanden Biesen, who later on the phone describes for me what I would characterize as a lifelong death fetish. On visits to his grandparents' home in Germany, he says he grew obsessed with the numerous crucifixes that adorned the walls. He lay awake nights thinking about death symbols and what happens in the afterlife. He was five at the time.

The fixation continued into adolescence. "I would sit and think about death and dying, and I tried to comprehend infinite nonexistence," he says. "I was consumed with it." He was probably the first Goth in Oak Park, Illinois, and he didn't even know it.

All signs to the contrary, Vanden Biesen did not grow up to become a serial killer. Instead he became a corporate pilot. One day in 1999, he was trolling around the Internet when he came across a periodic table. It was then he experienced his eureka moment.

"I looked at the elements, and I realized diamonds are made from carbon, and people are made from carbon." The notion struck him: If the carbon could be extracted from human remains, could they be turned by artificial process into real diamonds?

The idea drove him mad with excitement. "I was going berserk, talking to everybody about this," he says. For Vanden Biesen, the prospect eased his fears of death: "It was a sense of relief that there is something, that I could be something other than buried or cremated and forgotten. Something that people would cherish and remember and keep, that my memory could continue to live."

Despite his utter lack of scientific expertise, Vanden Biesen, who has a bachelor's in business administration from Cardinal Stritch College in Milwaukee, threw himself into his pursuit. He read about lab-created diamonds, about applying high pressure under high temperatures to create seed crystals, the same process that would apply in nature but at warp speed. "In general terms, I was convinced this was something we could do," he says.

At this point in his story, I should say that I am finding all of this improbable. I have written about business for many years, and never have I heard such a fantastic start-up tale. It should not be this easy. I mean, I have noticed that my basset hound's drool has roughly the same properties as John Frieda Frizz-Ease Hair Serum, and you don't see me incorporating.

Vanden Biesen insists his ignorance was an asset. "A lot of people have so much knowledge that it's a barrier to discovering something new," he says. "I wasn't tainted that way. I didn't know I couldn't do it."

Of course, Vanden Biesen had help. "It was up to my partners to find a commercially viable way to do it." Vanden Biesen had gabbed about his grand idea to his brother, Dean, who is married to Greg Herro's wife's sister. Instead of laughing at him, the men and Greg's brother, Mike, began discussing the concept at family gatherings, at first abstractly, then with growing excitement. Rusty supplied the research; Dean, a production manager for a company that sells steel parts, had the manufacturing experience; Mike could set up the order and tracking process; Greg, a computer consultant and entrepreneur, knew how to start and run a business.

One by one, all quit their jobs and maxed out their credit cards. In October 2004, Rusty Vanden Biesen stopped flying planes. "I told my wife we may have to sell the house," he says.

Meanwhile, they had settled on a way to make diamonds out of cremains. Greg Herro explains the process to me.

The first step is to collect the cremains. Eight ounces of ashes can yield enough carbon to make up to ten diamonds of up to one carat each in size. Eight ounces is only a smidgen of the four to six pounds of ashes a human body produces—all of which is enough, the company says, for at least one hundred diamonds.

The ashes are mailed by the funeral home to the company in Chicago, where they are placed in what is called a crucible. I picture something religious and a little bit ominous, but it turns out to be a metal tube about eight inches in height and four inches in radius. (Crucible, when I bother to look it up, can mean a severe test or trial but also means "a vessel made of a refractory substance such as graphite or porcelain, used for melting and calcining materials at high temperatures.") The crucible is then sealed and etched with a sixteen-digit number used to track the gem's journey.

The second step, which LifeGem calls "purification," is the one in which they capture the carbon from the cremains. The capturing occurs in a factory outside of Pittsburgh, in an industrial-size oven set at roughly 3,500 degrees Celsius. It takes one month, at the end of which you're left with pure graphite from the carbon. It looks like pencil lead—shiny, silvery particles. "Beautiful, really," says Herro.

In the third stage, LifeGem "grows" the diamond. The pencil-lead graphite is placed in an industrial press, a square machine that's eight feet in height, width, and depth—about the size and shape of the back half of a Hummer. This machine presses down on the stuff from every direction at extremely high temperatures for anywhere from three to seven weeks. The longer the press time, the larger the diamond.

Once the press is opened, out comes a diamond in the rough: a dark yellow or blue hunk of rock candy.

It is in the fourth stage that the diamond appears. The rock candy is sent to cutters in Bismarck, Nebraska. This is interesting to me. I had not realized Bismarck had a thriving economy in diamond cutting. "I don't know, that's where the best guys we found are," says Herro. They need the best:

Cutting diamonds made from cremains is very, very tricky. "They can crack," Herro says. "All diamonds can crack."

This, too, is interesting to me. I had not realized that diamonds, all diamonds, could crack; I had thought the diamond was the hardest substance on earth. I think about all the times I have whacked this little rock on my left ring finger on some unforgiving surface. As bad as I would feel cracking the stone my husband paid for by teaching fifty hours of clarinet lessons to fourth-graders, I think about how much worse I would feel if this stone were, say, my mom.

In order to ward against cracking and other problems, the cutter spends at least a month shaping and faceting the diamond. LifeGem's Web site says I can order my diamond in a round, princess, or radiant (a kind of rectangle) shape. If I have ordered a half-carat princess but they wind up with enough usable rock for a three-quarter-carat round, I can have that for no extra charge. Most folks choose the bigger rock.

Then, if I want, the cutter will laser-etch a message on the girdle, which is the fine edge around the abdomen of the diamond. No one can see the message without a loupe.

Finally, a certified gemologist from the Gemological Institute of America will check it out, labeling the rock an actual, honest-to-goodness diamond. This is a common misperception about LifeGem diamonds: Though the process to create them is artificial, the gem itself is not. The diamond receives a grade for size and brilliance, but not for color or clarity, given that only white diamonds receive those grades.

All LifeGem diamonds come in either yellow or blue. That was not necessarily by design—they just turned out that way—but LifeGem is trying to make the best of it, working hard to educate skeptical customers of the rarity and beauty of what are called, in industry parlance, "fancy" diamonds.

The yellow is intensely yellow with a spike of orange, like a cat's eye or a summer ale or, as the Web site says, "a sunset captured in time." The blue is prettier, in my opinion—like "a wave upon the ocean"—but costs twice as much because the process involves working with boron (a metalloid element) and takes some manipulation to achieve.

The result is breathtaking, as is the price. LifeGem charges $2,500 to $14,000 for diamonds that range in size from a quarter to one full carat.

"We hope to bring down the price as the technology improves," says Herro, sounding apologetic. Where once they hoped to see orders for one hundred diamonds a year, LifeGem now is making one hundred a month. I figure that amounts to a gross revenue of anywhere from $3 million to $16.8 million a year. I am thinking maybe Rusty Vanden Biesen is not so mental after all.

Back then, back in August 2002 when the company unveiled its product, all the world thought the whole thing was mental. WEIRD BUT TRUE! shrieked the *New York Post.* The *Maryland Gazette* included LifeGem diamonds in its list of "Wild and Wacky" holiday gifts. HERE'S A WAY TO MAKE A LASTING IM-PRESSION, sniggered the *Philadelphia Inquirer.* The *Orange County Register* called anyone who'd consider the process "looney-tooney survivors."

Imagine, then, being the first customer, the first person who couldn't care less if someone thought him looney-tooney, who felt this would be the best way to remember someone he loved very, very much. Imagine letting your loved one—no, *requesting* that she be the first human being ever turned into a diamond.

I couldn't. So I gave LifeGem's first customer a call.

VALERIE SEFTON
1974–2002

*B*ill Sefton is driving on an Illinois interstate when he takes my call. He owns a company in Chicago that processes credit card payments for banks, as well as a travel agency through which card owners can claim mileage. He travels to Chicago to conduct business, though his home is a five-acre ranch in Scottsdale, Arizona, where he lives with his wife, five horses, and forty vintage muscle cars. The cars—Camaros, Corvettes, and Dodge Darts from the 1960s and 1970s—live in a 5,500-square-foot garage with a giant mural of an old-time Dodge dealership. He calls his home the Red 'Vette Ranch.

Bill is funny and easygoing in the manner of someone who is used to success. But even people who usually win occasionally lose, and it is not so long ago that Bill Sefton lost what he loved most.

Valerie, Bill's daughter, died at age twenty-seven in the summer of 2002. She was diagnosed with Hodgkin's disease soon after she had graduated from Northern Illinois University in DeKalb. Valerie had just been licensed to teach K–3 at an elementary school.

Over four and a half years, Valerie endured ten bouts of chemotherapy and two stem cell transplants. Before her last transplant, she wrote a letter to her family. She told them where the letter was and asked that they open it if she didn't make it.

The second transplant seemed to go well; doctors said it was technically a success. But the new immune system caused other complications, including viral encephalitis, a swelling of the brain.

In those last days, Valerie lost track of time and place, waking from trips to Australia and other places she'd never been, delighting over and over again at the news of her sister's second pregnancy, dipping her burger in chocolate pudding and pronouncing the taste "terrible."

In those last days, Bill camped out at his business partner's coach house in Naperville, traveling to and from the University of Chicago Hospital, logging fewer than twenty hours a week at work, sharing bedside vigil with his ex-wife, Valerie's mom. One morning he opened the *Chicago Tribune* to an article about LifeGem. "I thought, Wow, this is a perfect Val kind of thing. This would be a cool thing to do," he says. He longed to show Valerie, but by this time she did not know her surroundings. He tucked the article away and didn't mention it to anyone.

Bill left the hospital at 7:00 p.m. the night of September 10, 2002, satisfied that Valerie's condition seemed stable. His ex-wife, Chris, a registered nurse, was about to leave, too, when Valerie's blood pressure began to fluctuate. Chris called Bill at various points to update him on Valerie's condition. No one was very worried.

At 11:00 p.m., Valerie died.

In their devastation, the family found and read Valerie's last letter. In it, she explained—calmly, decisively, and very specifically—what she wanted done after her death.

She would have no viewing of her body. The steroids and chemo had left her swollen and bald, and no young woman wants to be remembered that way. She would allow in its place a memorial service displaying her beloved scrapbooks.

Valerie was especially clear about her remains. She was to be cremated. "But if you put me up on a mantel somewhere, I will come back and haunt you," she wrote. "Take my ashes, and break them up into small lots. Give them to my friends and my family. Have them put them where they'd like me to be."

After reading the letter, Bill went and fetched the article on LifeGem. He showed it to the family. "Everyone agreed," he says, "that it was exactly what she would have wanted."

Bill called LifeGem and ordered six stones: for himself; Valerie's mother, Chris; his wife, Becky; his sister, Kathy Rinna; Valerie's sister, Tracey; and Tracey's daughter, Emma. He recalls they cost him between $16,000 and $18,000—a small sum, he says, compared with the funeral and all the other death costs. The smallest diamond, for Emma, was 0.25 carat; the largest, his, measures 0.55. All are blue.

Greg Herro flew out to Scottsdale to deliver the diamonds himself. It was then that Bill learned of Valerie's place in history. He had agreed to media coverage of the diamond's delivery, and the intense attention made him ask Herro: "Valerie's got to be among the first, right?"

Bill says Herro looked nervous. "She *is* the first," said Herro.

"Wow! How cool is that?" Bill says he responded. "My Valerie's the very first person ever turned into a diamond."

Imagine Herro's relief: LifeGem had lucked into a dream customer in Bill Sefton. He even gushes about the gem's imperfections. "It's unique because it's Valerie," he says. "The flaws in the stone give it a lot of fire and personality. That's Valerie, too. If I wanted a perfect stone, I'd go buy one."

Bill had his stone set into his wedding band, which he never removes. But the memorializing of Valerie Sefton did not end there. The diamonds fulfilled the spirit of Valerie's wishes, but the abundance of leftover ashes allowed the family to match the letter, too. They divvied them up in Ziploc bags. Her mother, Chris, found small wooden boxes with angels on top—Valerie had always loved angels—to act as mini-urns. Everywhere they went for the next few years, the angel boxes went, too. There's a little bit of Valerie in the waterfalls of Maui. There's some of her in Disneyland near a bronze statue of a little girl with a sea turtle (Valerie loved turtles). The workers at Bill's company took their box to her favorite place for lunch and left a little of her at the Cracker Barrel.

One portion of Valerie's ashes were set aside for Cori Ullman, her best friend. "You're a diamond in a pile of rocks," Valerie used to tell her. Cori couldn't afford a diamond but swore she would someday. Last year, Cori got married. As a wedding gift, the Seftons gave Cori a canary yellow 0.6-carat diamond made of her best friend. She could not be the maid of honor, but Valerie was right there.

Bill is quiet for a moment. "Most times, when I think of her, I smile," he says, and his voice breaks, and he cries.

I picture this vibrant, generous man, driving alone down an Illinois highway, sobbing as he tells a stranger about the daughter he lost, the daughter who now sparkles from his left ring finger. It is not Valerie, but it is part of her, and that is something.

When he can speak again, Bill does not apologize for crying. He lost his girl. He can remember her however he damn well wants. He can cry before a stranger if he damn well wants. I am crying, too.

"We went through some real bad stuff for a real long time," he says. "There's not a lot of good stuff to remember. Every day I look at my ring. Every day I appreciate it. It's one of the most fulfilling things I've ever done."

❖

SCENE FIVE: The Homecoming. In a high-ceilinged brownstone apartment in Manhattan's East Village, two women sit at a large, thick-legged dining room table. Between them on the table is a FedEx package, about the size of but much lighter than a dictionary. PEGGY ATKINSON, the woman dressed all in black sitting at the head of the table, nervously adjusts her glasses and smooths her ash blond wig. An expression of determination crosses her face. She takes a deep, rattly breath.

PEGGY: Well. Shall we? (Peggy, small and weakened by the cancer treatments, struggles with the box. The other woman, a JOURNALIST, takes it from her and carefully wrests open the gluey parts. There's a box inside a box, and then another box, like a Russian doll.)

PEGGY (murmurs): They don't make it easy, do they. (Finally they get to the bubble wrap, which surrounds a square black box about the size and weight of a drink coaster. Hands trembling, Peggy takes the package. It is a beautiful wooden jewelry box with a glass top. There, just under the glass on a velvety black surface, is DON—Don Atkinson, the diamond. The gem is round, the cut brilliant, the clarity high, the color bright yellow, the weight 0.5 carat.)

PEGGY (whispers): Oh, for God's sake, look at that. (Slowly, she opens the glass lid and rolls the gem gently into the palm of her hand.)

PEGGY: Oh, God. Isn't that beautiful? Look at it! (Suddenly she is consumed by the enormity of the moment. Here he is, her beloved husband, glinting up at her with all the sparkle and fire she remembered. Behind her glasses, her wide eyes brighten with tears.)

PEGGY: Oh, Donny. Donny. (When she speaks, she marvels in a steady, low voice about the wonder of the thing.) It's a half carat—all I could afford right now—but they said it came out a little bigger than that . . . I didn't know how the yellowing would turn out . . . have you seen others? Are they all as pretty as this? God, it's gorgeous . . . know what's engraved on it? "Shine on." That's what it says.

JOURNALIST: What will you do with it?

PEGGY: Oh, I've thought about it. Don designed our gold wedding bands. See—wavy, hand-chased. After his death I wore his on a chain around my neck, until it got too heavy—physically and emotionally. I'll go to one of Don's jewelers. I'll ask him to make his wedding band and the diamond into a pendant. Then I'll wear it. Always. (She takes one long look before she returns the diamond to the box.)

PEGGY: I'm glad I did this. I know this is what he would have wanted. (She exhales, looks far away. Smiles. The moment is bittersweet.) I'd much rather have my husband, but . . .

❖

As Near to Heaven By Sea as by Land

TWO WAYS TO SLEEP WITH THE FISHES

Families watch as artificial reefs, mixed with their loved ones' cremated remains, are "placed" in the waters off of Ocean City, New Jersey.

I like the ocean. I do. My hometown in Japan slopes gently to the Pacific, upon which, on a clear day, you can see the island where my mother was born. We couldn't swim in those waters—Kobe is a major port city—but growing up, I spent every summer where my American father had, at the Jersey shore. His family's summer house faced the beach, and every day my siblings raced across the sand to fly flailing into the seaweed-green waves. I stayed on the porch. Sometimes I waded in, but, truth to tell, even on its calmest days the latent volatility of the Atlantic frightened me. Plus, ocean swimming is kind of messy.

I like the ocean, I do—but I wouldn't want to be buried there.

Many people would. Of these people, the sea commands a love so deep and fierce as to define them. These are the people for whom souvenir shops on the boardwalk stock tchotchkes involving seashells and talking bass. These are the people who body-surf and skin-dive and deep-sea-fish and polar-bear-dip. These are the people who can envision nothing more peaceful than to spend their eternity at sea.

So many people desire a disposition at sea that a cottage industry has bubbled up to accommodate their wishes. Bear in mind, a whole-body burial at sea is still not quite a snap to arrange, unless one was or is in the navy. But the popularity of cremation has allowed for a head-spinning array of creative options from merchants, many of them baby boomers themselves, by which one can rest with the fishes.

At the National Funeral Directors Association convention in Nashville, I met a pair of gentlemen who were selling biodegradable urns shaped like giant seashells the size of toilet lids. When tossed onto the water, they float awhile before sinking and eventually dissolving. In 2004, a Miami business-man announced plans for an underwater cemetery for cremains, modeled after the lost city of Atlantis. Charter boats that usually host weddings and deep-sea-fishing expeditions now advertise their services in mortuary mag-azines and on obituary pages.

Of course, the practice of ash disposition at sea is hardly new. I venture to guess that many a family has gathered on a beach at dusk past the empty lifeguard stands to cast Pop-pop's cremains into the surf. They are breaking the law. The Environmental Protection Agency rules that human cremated remains must be discarded three miles from shore, no closer. I imagine this law is not easy to enforce. I personally have never seen federal agents skulk-ing around the Jersey shore at twilight, arresting sad-looking families carry-ing bags of dust.

Even so, most people, when given the option, prefer to stay on the cor-rect side of the law. Plus, ocean scattering is kind of messy. You want to comply with Pop-pop's wishes, you do—but you just wish someone would do it for you. These merchants simply take what is probably one of the oldest forms of death personalization and handle the logistics.

EPA guidelines are a tad stricter when it comes to whole-body burials. The agency stipulates that "burial at sea of human remains that are not cremated shall take place at least three nautical miles from land and in water at least 600 feet deep"—a rule I appreciate, as it assures me that, with all of my ocean-related phobias, dog-paddling into a corpse need not be one of them.

The EPA adds that "all necessary measures shall be taken to ensure that the remains sink to the bottom rapidly and permanently." On a form that must be filed with the EPA within thirty days of sea burial, one must answer the question, "Did the remains appear to rapidly sink to the ocean floor? Yes/No."

Curious about how exactly this was achieved, I consult the United States Navy guidelines for at-sea burials.* Not surprisingly, burial at sea is a popular form of disposition for members of that military branch. In its guidelines, diagrams show where to place a minimum of six nylon or metal bands to keep the casket from opening underwater. It must also be notched with twenty holes, each two inches in diameter. Finally, a 150-pound weight at the bottom end of the casket "ensures 'feet-first' sinking."

The logistics may be easier to navigate today, but the at-sea burial of cremains, as of corpses, still brings up concerns not met in land burials. Religion, for one. The Catholic Church prefers land burials but will accept burials at sea, which, in the case of cremains, means sinking a sealed urn toward the ocean bottom. John F. Kennedy Jr., a Catholic and the son of a navy man, received a navy burial of his cremains off the shores of Cape Cod. Hinduism approves of scattering cremains in a specific body of water: the Ganges River. Muslims are permitted at-sea burials only if land burials aren't possible—say, in the case of a sailor who dies at sea. Jews are advised to consult first with a rabbi.

Here's another issue. There's no going back when you're buried at sea—no exhumation, no relocation to that fancy new columbarium up the street.

---

* U.S. Navy Mortuary Affairs, Burial at Sea Program brochure.

Plus, there's no going back for the family to visit the burial plot—no annual picnic by Pop-pop's tombstone, at least not without scuba gear.

On the other hand, when you're buried in the ocean, two-thirds of the earth is, arguably, a memorial to you.

On a weekend in May when the Atlantic glitters invitingly, I can see how that prospect would appeal. So on said weekend in May, I tagged along to witness two very different ocean farewells: one from the sky, the other to the bottom of the sea.

• PART ONE: SCATTERING ASHES BY AIR •

Flying seems to be one of those activities that can't be categorized as either work or love. By day, Bill Fallon pilots jet airplanes for American Airlines. At forty-three, he is a twenty-year veteran, a captain since 1999. To me, a frequent flier but a nonpilot, this seems like a stressful and tiring job. Yet on his days off, Bill packs his thick-chested frame into a Cessna 182 single-engine four-seater.

"You kinda get worn out flying big planes—it takes all the fun out of it," he says. "I like little planes, little airports. Someday I want to teach my daughter to fly in this."

Bill Fallon's Cessna 182 serves three main purposes: thrilling his four-year-old daughter with joyrides, ferrying friends to Atlantic City, and scattering people's cremated remains. Today's flight is neither a joyride, though the day is sunny and warm, nor a gambling jaunt, though at one point Bill realizes he has $400 and could make a quick casino run.

Today, we are on what he calls a "mission."

We are scattering the ashes of Arthur Jensen and Algirdis Jovais, who are the husband and the brother, respectively, of one Donna Jensen. The family is gathering on the beach in the town of Beach Haven, New Jersey, about one hundred miles south of New York City, three nautical miles from where we will make the drop. I say "we" because Bill needs assistance working his homemade ash-scattering device, and I am the only other living person accompanying him on this flight.

I don't know many airline pilots, but I have met a lot of funeral directors, and Bill reminds me of the best of them: He's kind and personable and solicitous to a fault. He flies out of his way to pick me up at Teterboro Airport to save me an hour's drive. He fetches me bottled water before we go and consults a worn flight map to plan the most scenic route. He flies over my house.

As we buzz down the Hudson River past the Statue of Liberty, Bill tells me how his side job in ash scattering came about. We are wearing headsets with microphones to hear each other over the engine, which gives our conversation a walkie-talkie effect. He bought his gray-striped, 1976 four-seater for $68,000 as a plaything just before September 11, 2001. A training buddy of Bill's was a pilot on Flight 77, which crashed into the Pentagon that day. Aside from the shock and horror and grief, Bill and his colleagues suffered lasting consequences: As the industry reeled and they fought to save their company, American Airlines employees took big pay cuts.

With flight times slashed, too, Bill's income swooned from $200,000 to $138,000. That's still a handsome living, Bill admits, but with a family to support, it's not really enough to justify a toy that costs $100 an hour to operate. The Cessna, said his wife, Maryann, would have to pay for itself.

One day, a hearse pulled up to the airport in northern New Jersey where Bill parks his plane. The driver got out, looking lost. "What do you need?" Bill asked. The driver was a funeral director, and the box he was carrying held the ashes of a man whose last wish was to be scattered over the Watchung Mountains.

"Well," said Bill, "I can do that for you."

He checked with the Federal Aviation Administration and the Environmental Protection Agency and could find no law preventing him from spreading ashes from the air above public land, as long as he abided by local littering laws and obtained the proper permits.

Navigating the flight restrictions on pilots was a little trickier. Commercial pilots are allowed only so many hours per month of paid flight, according to part 135 of the FAA's rules, says Bill. Crop dusting, however, falls under part 91, a different category. Bill proposed to the FAA that his business was in fact a form of crop dusting, which, if true, would allow him a side business in scattering without restricting his jet-flying hours.

Maybe the FAA had bigger fish to fry just then, but anyway, it worked. In 2002, Last Wish, Inc., was born.

Bill placed ads in the Wednesday obituary section of north Jersey newspapers and launched a Web site featuring organ music and a picture of a plane flying into the sunset. "Not an ending," reads the slogan, "but a dignified departure on a new journey." His charge: $350 a drop, which with fuel costs, landing fees, and marketing, means he just about breaks even.

Most of the requests are for sea scatterings, but occasionally families request ash disposition over certain landmarks. Bill contacts the township, though he has learned from experience not to ask point-blank for permission. "If you say, 'Hey, can I drop ashes over your town?'—of course the first response is no," he says. Instead, he inquires if there is a specific law on the books prohibiting ash scattering. Usually there isn't, and he proceeds, his conscience clear.

Once a man's family asked for his ashes to be scattered over his favorite golf course. The club knew him well and allowed it, though Bill made the drop at night when the remains were unlikely to dust a duffer. Another time, he got a request for a scattering over the deceased's favorite bar. That one didn't fly.

———

The little plane turns inland until we spot the Eagle's Nest landing strip, where we are to meet Donna Jensen. It's literally a strip of asphalt cut into a forest near a quarry. Bill touches down. He frets over the bumpy landing.

"I hope the family didn't see that. I'd be so embarrassed." We pull up near a trailer that appears to serve as an office, next to which is parked a black SUV.

A white-haired woman gets out. She's tanned and wearing big sunglasses, a white suit, and a coral-colored top. A boy hops out of the back.

Bill strides off the tarmac toward the SUV and grasps Donna Jensen's hand with both of his. He calls her Mrs. Jensen. He tells her how he will conduct the scattering, adding that after he is done, he will fly west, as they do in military formations to signify the death of a comrade.

"Oh, Arthur would have loved that," she says, nodding enthusiastically. "He was into all that stuff."

The boy is named Mischa, and he is Donna's grandson. Mischa was adopted by her daughter from Russia, she tells us, which is perhaps why at ten he is still unjaded.

"I'll be waving Spidey Bear when you fly over," he tells us, holding up a teddy bear dressed in a Spider-Man costume. "Look for him. Okay?"

We head back to the plane. Bill lets Mischa climb into the cockpit, striking the boy speechless with wonder. Meanwhile, Donna reaches into the Lands' End tote bag Mischa had lugged over and pulls out two urns.

They are black and made of plastic, the size and shape of shoeboxes, their contents surprisingly heavy.

ARTHUR JENSEN
1921–2004

ALGIRDIS JOVAIS
1921–2004

*W*hy is it that death so often happens in threes? In the fall of 2004, Donna Jensen lost her husband of fifty-eight years, her brother, and her husband's cherished aunt. The aunt went last. In a New Jersey motel room on the night of the aunt's funeral, Donna picked up a local newspaper to scan the obituaries. Amid the death notices she saw an ad. "Not an ending . . . but a dignified departure on a new journey," it read, with a picture of a plane.

That's it, she thought.

Donna Jovais was born and raised in Brooklyn, the daughter of a Lithuanian immigrant. As the war began, Donna took a job at Guarantee Trust, a bank in downtown New York City. One day, a former employee came back to show off his brand-new officer's uniform. The Brooklyn girl and the Jersey City boy fell in love—just as Arthur Jensen was to report for duty. Arthur wrote her from his station in the

Philippines, comforting her with the knowledge that as an army engineer, he was a step removed from combat.

He married her the instant he returned, in 1946, never to leave her side again. Arthur chose to attend the University of Maryland on the GI Bill because it was the rare school with housing for families (Donna gave birth to their first daughter there). Though a born-and-bred Jersey boy, he soon had U of M's red, white, black, and gold running through his veins. But family ties are strong, and after graduation he moved his family back to his home state for an executive job at AT&T.

Arthur was a solid man, unshakably loyal to the things and people he loved. He loved his wife. He loved his daughters, Gail and Robin. He loved the Jersey shore, where he spent nearly every summer of his life. He loved the army. He loved the University of Maryland Terrapins. He remained with AT&T his entire career.

Alzheimer's did not take him down without a fight. It was years before people could even sense a strangeness. Then one day—during a physical therapy session at a rehab center—the disease knocked him down for good. Arthur fell. He was rushed to brain surgery. He would never walk again.

The girls, both of whom attended Arthur's alma mater, now pleaded with Donna to move closer to their homes down south. She sold their home in Glen Rock, New Jersey. Arthur moved into an assisted living facility in Ellicott City, Maryland, Donna to a condo nearby.

"After that," says Donna, "he didn't last long."

On November 11, 2004, Arthur Jensen died.

They had spoken, back when he could, about the end—and thereafter. He wanted to be cremated. So did she. He wanted his ashes scattered over the ocean at the Sixteenth Street beach in Beach Haven, New Jersey, where he had spent nearly every summer of his life. So did she. They wrote it down on paper and signed it together.

Just two months before, Donna had learned that her brother Algirdis Jovais had also died. Algirdis had seven children from a previous marriage, most of whom she says were estranged. He had remarried, and his wife had a family burial plot back in Germany, but it was full up—and in any case, he had once spoken of his wish to be buried at sea, if not in Lithuania.

As she sat in the motel room looking at the obituary page, Donna decided she would have her husband's and her brother's ashes scattered together over the ocean they both loved. "This way, I thought, Well, maybe some of my brother's ashes will drift all the way across the ocean. Who knows? It's a nice thought, anyway."

Donna called the number in the ad. Bill Fallon answered. They planned a date the following May when Arthur's relatives would gather at the Sixteenth Street beach. They decided on a meeting place to hand off the ashes. Donna's friend Bob, a colleague from her days teaching school in Glen Rock, would play his bagpipes. She would say her final good-byes to Arthur and Algirdis.

❖

As Mrs. Jensen drives away, Bill shows me his ash-scattering apparatus. I had in fact been wondering about the how of spraying cremains from a plane; even with two people, I could not imagine pulling it off in a neat and orderly manner.

At the back end of the cabin is a hatch, beside which Bill has rigged what looks like a wooden stepstool. White PVC piping sticks straight up out of it. The bottom end of the pipe is curved so that it can poke out of the open hatch.

The apparatus is really quite simple. The ashes go in through the top of the pipe and shoot out the other end, sort of like an upside-down chimney. Once the ashes are poured in, a lid is secured and a gate valve keeps the cremains from flying out until it's time. To this homemade contraption Bill has added a recent innovation: a tank filled with pressurized air, attached by nozzle to the pipe, to force out every last bit of the cremains.

Bill has not yet patented his invention, but when he does, I decide he should call it the Scatterrific.

As he secures the open hatch with some bungee cord to keep it from flapping, Bill explains my role in operating the Scatterrific. First, I am to pull the lever on the gate valve. This opens the chute and releases the initial batch. Then I twist the knob on the air tank to flush out the rest. He shows me what happens if I forget to pull the lever first: The pressurized air causes the cap on the chimney pipe to pop off.

"That would be bad," says Bill.

I picture the cabin suddenly clouded with the remains of Mr. Jensen, the plane veering wildly over the Atlantic as Bill struggles to breathe and see. The Jensen family sues Bill, and I have to shower for hours.

It would be bad.

This is the first time Bill is scattering two people's remains at once. The Scatterrific could probably hold the contents of both urns, but Bill insists on separate scatterings, even though he billed Mrs. Jensen for only one. (Many end-trepreneurs, I am learning, lack a business gene.)

No matter: Out comes the backup scatterer. This one he bought on the Internet from a Colorado outfit called Trail's End that boasted its product was the only one designed specifically for the task. It's a long nylon sock, open on one end and clamped on the other with a leather handle. You're meant to pour the ashes into the open end, roll it up, and secure it with the leather strings. The pilot holds it out his window and undoes the string—and voilà, a scattering.

"Four hundred bucks for this," Bill mutters.

Bill opens the plastic urns and pulls out clear plastic bags sealed with twist ties. The ashes are white gray. A body does not emerge from the incinerator this way; at a New Jersey crematorium I visited, I watched as the remains were poured into what looked like a milkshake maker and pulverized to dust. "Sometimes," Bill says later, "the crematorium doesn't do its job and you get unpulverized bone the size of golf balls that ding the side of the plane."

Right now he is silent with concentration as he carefully pours Arthur Jensen's remains into the Scatterrific; Algirdis Jovais's go into the Trail's End bag. Bill works inside the cabin to shield himself against the wind. I wait in the backseat.

Outside, a crowd has gathered. I had noticed when we touched ground that a clump of men were sitting in lawn chairs by some pickup trucks near the airstrip. They appeared to be flying some model planes.

One had ambled over after we landed, saying they were a flying club. No, he didn't fly-fly, like, airplane-fly. I have lived in New Jersey for many years now, and I am still surprised by the variety of accents. This guy's is straight from *King of the Hill*.

"Plane geeks," Bill whispers.

The flightless geek had apparently told his fellow geeks about Bill's mission, and they had clustered to gawk. One, it turns out, is a mortician in training. He takes Bill's card and promises to recommend his services. Bill shrugs. You never know.

At last, Bill's cell phone rings. The Jensens have arrived at the beach. They are gathered and ready.

<hr />

We take flight and head toward the ocean. On the beach, we spot a large group with two dogs. I squint but can't see Mischa's Spidey Bear. A couple of people wave. Bill dips his wings to show he sees them. Then he flies directly over them and heads straight out to sea.

The Atlantic unfolds before us. Some boaters are on the water, taking advantage of the strong sun and warm winds. We keep going at eighty knots for what seems a long time. I wonder if Mrs. Jensen and her family can even see us anymore. Later, she tells me she could not.

Finally, Bill turns his Cessna around. "Ready?" he says to me.

I am nervous. What was the order again?

Bill helps. "Valve first," he says. I pull the black lever. Outside the left rear window, I see a dusty cloud as Arthur Jensen bursts out over the water. His particles twinkle in the sunlight.

"Pump next," says Bill. I twist the lever on the tank, and the pressurized air makes a *whoosh* as the rest of Mr. Jensen leaves the plane and takes flight. I watch him billow on the wind as far as I can see. It's beautiful.

When Mr. Jensen has left the plane, Bill opens his window and holds out the nylon bag by its leather handle. He steers with his knees as he struggles to undo the leather strings.

I have seen this move. My husband does this wielding a Whopper in one hand and a Coke in the other while doing seventy on the freeway. It doesn't feel particularly safe one thousand feet over the ocean, either.

Bill has tied the leather strings so tightly, they now are stuck. He growls something I don't catch.

Finally, the bag unfurls. Algirdis Jovais escapes from the plane and on to the windstream toward Lithuania. Who knows? It's a nice thought, anyway.

The plane buzzes in a straight line back toward shore. Bill doesn't speak. His face is solemn as we soar over the family. He wags his wings and flies west, following the path of the sun.

• PART TWO: THE ARTIFICIAL REEF AS GRAVE •

It is a humid and drizzly Sunday. There is a crowd at the Shamrock Marina in Somers Point, near the southernmost tip of New Jersey. The marina is long and narrow, one side lined with leisure boats hoisted up on stocks and the other side with a row of what look like giant concrete beehives. There are nine of these beehives, about waist high, each punched with holes the size of volleyballs. People mill around wearing shorts and baseball caps, laughing and crying and posing for pictures. Children scratch at the rough surfaces with chalk, blotting out the gray of the concrete with pink and yellow and sky blue.

The beehives contain cremains, the happy-sad people are the bereaved, and it is the strangest viewing I've ever attended.

The viewing has been arranged by Eternal Reefs, a Decatur, Georgia, company that offers burials at sea within artificial reefs. Let me explain. Each of the beehive sculptures—the company calls them "reef balls"—contains one person's cremated remains mixed into the concrete. A six-inch bronze plaque proclaims the deceased's name and dates. The reef balls are then "placed" in the ocean during an at-sea dedication ceremony. The balls become part of a state-sanctioned reef-rebuilding project and within months host an active community of fish and wildlife.

When I tell people about the reef balls, I get mixed reactions. My sister Emy, who like me has spent almost every summer of her life not a mile from Somers Point, exclaims, "I'm swimming with dead people?!"

My cousin Greg has a different take on at-sea burials in concrete. He points out: "The Mob has been doing that a lot longer—and for free."

Eternal Reefs is often the kicker in newspaper articles about funeral trends, inebriating headline writers with a minibar of cliché opportunities ("sleeping with the fishes," "deep-sixing your ashes," and so on). I, too, am guilty of this; I remember a TV appearance after my article ran in *Time* in which I used this company as a punch line.

I had not at that point experienced a placement myself. So I was entirely unprepared for the emotional impact of the viewing and burial of what are, after all, big blobs of concrete.

Don Brawley says he was unprepared, too. Not long after graduating from the University of Georgia, Brawley, an environmentally minded scuba enthusiast, had started a business casting artificial reefs out of concrete for state-sponsored reef-rebuilding projects. One evening over dinner, his father-in-law asked to have his remains buried at sea in one of them.

"Sure," Don remembers saying.

A few months later, his father-in-law died. Don scrambled to find the regulatory loophole that would allow him to do as he had been asked. He did: Cremated remains, once mixed with concrete, are no longer considered remains; they are a concrete additive. He founded Eternal Reefs in 1999.

I meet Brawley, forty-two, at the viewing in Somers Point. He wears an Eternal Reefs polo shirt, speaks earnestly, and says things like, "We're in the closure business." With his red beard and pale skin, I would not have pegged him a scuba bum.

"We realized there were a lot of people who say, 'Oh, scatter me,' but their families didn't like that—they felt they were throwing their loved ones away," he is telling a group of curious family members.

The balls cost between $1,995 for the knee-high, four-hundred-pound Aquarius and $4,995 for the chest-high, four-ton Atlantis. Brawley says that almost immediately after placement, underwater critters take up residence. Within six months, it's a condo with no vacancies. In five years, it's indistinguishable from natural reefs, which take twenty to two hundred years to grow on their own.

As for the families, says Brawley, "What we're giving them is a living memorial. A place to go. A ceremony, a process."

That process begins in an industrial cement yard in Sarasota, Florida, where families are invited to participate in the reef ball casting. The cremains are poured into five-gallon buckets, into which two gallons of concrete are mixed. The cremain-concrete blend is added to the rest of the concrete as it's poured into the beehive mold. Once it's solidified but still wet, family members can squish their handprints or scratch final messages into the ball. One month later, the balls are ready and transported to the coasts of Florida, Georgia, South Carolina, New Jersey—"anywhere, really, with a reef-building program"—for the placement ceremony.

Nine people are to be honored in this weekend's ceremony, and thus nine separate blobs of concrete are to be placed in the sea. They include fishermen and deep-sea divers and at least one lady who loved the sea but could not swim.

Two of the nine are veterans, and, blob of concrete or not, every veteran in good standing merits a military funeral. Earlier in the day, two sets of men in uniform—navy and army—had fussed over two of the reef balls, arranging American flags. I had admired their seriousness as they puzzled over how exactly to fold the flags so they didn't touch the ground or cover the name plaques.

At noon, the crowd falls quiet. Three young navy officers in their white dress uniforms slowly approach one reef ball. Two face the ball and salute. The third raises a bugle to his lips, and "Taps" echoes over the boatyard.

I am close enough to notice that the sailor isn't actually playing. With the demand for qualified buglers outstripping supply, the military is lately embedding computer chips in the instruments to offer at least the lip-synched version; this is not the first funeral I've attended with fake "Taps."

It doesn't matter. Everyone, including me, is wiping their eyes.

The sailors then step to either side of the reef ball, lean over to pick up the flag, fold it into a perfect triangle, and present it to the widow. The process is repeated by two army officers in their moss green dress uniforms. Afterward, the two widows stand facing their husbands' reef balls. Just as they do at Arlington, the widows press the folded flags to their chests and weep.

The next morning, the families converge at 7:30 at a dock in nearby Ocean City. Three boats are waiting, each to hold fifty people or so. Two hold a mix of families. The third boat, the *Miss Beach Haven*, is taken up entirely by the family and friends of John V. Slowe, who died of a heart attack while watching TV. His gravelly voiced widow is herding people onto the boat, greeting even the unexpected guests with warmth, even though each passenger is costing her an extra $50.

I am on the *Captain Collet* with six families. The humidity of yesterday has dissipated into a blue-lipped chill, and the air grows icier the farther we head out to sea. Some huddle inside the small cabin near the upright heater, rubbing their hands and looking green. Others sit on benches on the deck, braving the salty mist and the loud winds, eyes fixed on the horizon.

The boats chug for an hour out to the designated coordinates of the placement seven miles from shore. The exact location matters because the placements are part of New Jersey's official reef-building project. Also, the company gives families the coordinates so they may visit the site again. I wonder how many actually do; it's not exactly as easy as a Sunday drive to a cemetery. Still, even if they never see these dark waters again, I can understand that it would give a family comfort to know where in the great ocean a loved one lies.

KEN SIMPSON
1924–2004

*K*en Simpson was a Brooklyn boy, but he'd take the sea over the city any day. He lifeguarded on the beaches of Coney Island. He cofounded a polar-bear club that is still one hundred members strong. He spearfished. He dove sixty feet with no tank and called himself the 1955 skin-diving champ.

He met his future wife in the water. Ken had just returned from fighting in Europe. Mary was attending Long Island University to become a phys ed teacher, but first she needed to pass the swimming exam.

She didn't swim. At the pool, she found herself alone among a dozen boys who all offered to help Mary learn her strokes. Ken was polite. "Oh, and he was nice-looking, handsome," she says. No small thing.

They were an odd couple, but perhaps not: She could have a long conversation with a telephone pole, says their daughter, and one could say he was that pole. They married. Both became schoolteachers; he taught sixth grade, she taught fourth. They moved to California, but the waters of the Pacific were cold and uninteresting to Ken. They moved to Key West, Florida, where their daughter was born. They named her Mary Key.

Teaching didn't suit Ken, or rather, his job didn't suit his hobby. "What he really wanted was to be a beach bum," says Mary Key. But soon he landed on the ideal career. As a fire fighter, he could use water to save lives, plus the job's sporadic schedule—two days on, one day off—allowed him frequent trysts with his love.

His wife complained that she never saw him. "He was reticent and kept to himself, and the sea provided him a place to do that," says Mary Key. "You go underwater, and you can't hear anybody."

Unlike her mother, Mary Key loved her father's ocean. She tagged along on what she called their "adventures." They'd pack tuna-fish sandwiches and Pepsis and leave before dawn. They'd take their little boat out to Kerry's Fort Lighthouse off Cutler Beach. They'd climb to the top and eat their sandwiches and see fish in the clear water. They'd snorkel for hours.

Water is a fickle lover, and Ken Simpson knew this. A gifted spearfisherman, Ken used to troll the water pits in the rock quarries for a catch. Spearfishing, like many underwater sports, is a test of endurance, and Ken could hold his breath like few others. One friend was not as lucky. Mary Key remembers being driven one night to the edge of a quarry to watch as police boats searched for the body. Ken didn't quit diving, and his daughter didn't worry. After all, she says, "he was invincible to me."

She is now Mary Key Brents, fifty-two, a massage therapist with a chiropractor husband and two young daughters. In 2004, they celebrated Thanksgiving all together at her home in Atlanta. A series of small strokes had lapped like steady waves at her father, leaving him weak, corroded. Mary Key had recently come across an article about Eternal Reefs, and she broached the subject with him.

Ken Simpson laughed. "I'm going to live forever," he said.

"Let's get real," said his daughter. "We're all going, Daddy, and chances are you'll go before I will. So let's talk about this. You're not an earth man. How about the sea?"

"I like it," he said. "It's a good idea."

Mary Key was glad. "I had an inkling it would be the last time I saw him," she says.

It was. Before the year ended, Ken Simpson was gone.

Mary Key made the arrangements, and her mother, Mary, traveled to Sarasota for the casting. She pressed her handprints in Ken's reef ball. A month later, the two Marys had flown to New Jersey for the ceremony. Ken's relatives still lived in this region, and many had made it out for the viewing. Now a handful were aboard the *Captain Collet*, waiting for Ken to make his final dive.

❖

*T*he three boats carrying family members have come to a stop. They circle around a red towboat called the *Defiant* that is carrying the reef balls. People find their families, hold hands, and fall silent.

Aboard the towboat, the first reef ball—that of Eleanor Baldwin—is bound with rope, hooked up to a crane, and lifted onto wooden planks, which slant like a slide into the sea. Six men ease it down the plank. A woman standing near me blurts out a sob.

This is a considerable operation, as the ball weighs 1,500 pounds. The men struggle to unbind the ropes. The sobbing woman watches, her grief stuck. We all hold our breath. Finally, Eleanor Baldwin's reef ball slips into the water toward the ocean floor seventy-five feet below. The whole process takes about twenty minutes; there are eight more to go.

It is at this highly inopportune moment that I begin to feel sick. It has been an uncomfortable few days. I spent a portion of Saturday's Cessna ride clutching Bill Fallon's airsickness bag, and my arms are covered with itchy welts from the gnats that also paid their respects at the shipyard viewing yesterday. Five minutes into this morning's boat ride, I was quaking with cold and queasy from the rolling waves.

The sign on the restroom door in the cabin reads, "Do NOT get sick in here—be courteous to others and go outside!" But I'll be damned if a family has to say its last good-byes over the sound of my retching and the sight of my regurgitated blueberry muffin.

I throw up into the bowl. Then I throw up again.

I mention this mortifying episode to make this point: Modern funerals and burials aren't about the comfort of the living. Many, in fact, involve distinct and intense discomfort—from braving cold to battling insects to barfing on a boat seven miles at sea. Anyone can sit in an air-conditioned funeral home, dabbing at their eyes with a perfumed hankie, their only inconvenience a stiletto heel sinking into the turf at the burial site.

No, these families here are swimming that extra mile for the one they loved—layering on the sweaters, blasting the bug spray, popping Dramamine by the fistful. I forgot to do all three and must suffer the consequences.

It is two more hours until we arrive at the final placement, that of John V. Slowe, whose reef ball is a supersized Atlantis. When it finally submerges, the *Miss Beach Haven*, upon which fifty of his family and friends are gathered, erupts in cheers and tossed flowers.

Don Brawley's partner, George Frankel, takes a megaphone and reads the words of President John F. Kennedy:

> *I really don't know why it is that all of us are so committed to the sea . . .*
> *I think it's because we all came from the sea . . . and it is an interesting*
> *biological fact that all of us have in our veins the exact same percentage of*
> *salt in our blood that exists in the ocean. And therefore we have salt in*
> *our blood, in our sweat, and in our tears. We are tied to the ocean, and*
> *when we go back to the sea—whether it is to sail or to watch it—we are*
> *going back from whence we came.*

On the deck is a cluster of miniature reef balls, each the size of a helmet, one for each family. Mary Key Brents has stuck red carnations, orange lilies, and white and pink roses into theirs. Together with her mother, Mary, she drops the miniature replica into the sea. The flowers bob to the surface. Later, I see that these replicas can be ordered from the Web site for $20, to be placed "on the desk or in the home."

Another family, the Ropers, have also filled their replica with flowers. Kenneth C. Roper of Atlantic City loved fishing so much that he handmade fishing poles for each of his six daughters. Down in Sarasota, they had pressed a paw print in his reef ball to represent his springer spaniel, Murphy. They drop their replica in the ocean, shouting, "Go, Eagles!"

Janice, his widow, gazes after it. She has already paid for her own Eternal Reefs burial. Someday she will join him there at the ocean bottom.

Back on the docks before we boarded the boats, I had met a woman named Cecilia Bachinsky, sister-in-law of Ken Simpson. "This is just not for me," she had said. "I'm Catholic," she'd added meaningfully, as if it were objection enough—which, of course, it is. "I would want something traditional. You know, normal." She'd come along on the trip somewhat reluctantly.

As we splutter back toward shore, I join her as she sits on a bench, looking pensive. "My eyes are opened," she begins slowly. She feels unexpectedly moved, she says, by the placements. She still feels the lack of ceremony, that prayers could have been said as the reef balls were interred, the horns on the boats sounded—something.

But looking out over the water toward the horizon, she says she understands. A traditional funeral, a "normal" funeral, would not have brought her any closer to her seldom-seen, intensely private brother-in-law. This experience had. "Now, my children will always associate the ocean with Uncle Kenny."

We sit quietly as the docks come into sight. The sun is finally breaking through. When we reach land, the Roper women prepare to disembark. One of them is as green as her Eagles baseball cap.

"All I can say," she says to her mother, "is I better go before you. I'm not doing that again."

Outside the Box

"FANTASTIC AFTERLIFE VEHICLES"
AND OTHER COOL CONTAINERS FOR CORPSES

*Batesville casket company's traveling tour bus rolls into Saddle Brook,
New Jersey, where sales representative Meghan Lopatich shows off the latest designs.*

Caskets are big business.

This I know because the casket companies have by far the most over-the-top, spectacular displays at the National Funeral Directors Association convention in Nashville.

They're clustered in the back of the giant expo but impossible to miss, what with the spotlights and music and performance artists. There's a long-haired man with a French accent finger painting to "Proud to Be an American." Nearby is a troupe of Broadway-style singers costumed to represent the Working Man—at least, that's my best guess. The cook, waiter, janitor, et al. end their routine belting "Circle of Life" from *The Lion King*. Maybe the point is that folks who work for a living eventually end up in a casket?

Demographically speaking, the future looks bright for the $3.5 billion casket industry: Over the next twenty years, as baby boomers age, the death rate will swell from 2.3 million a year to 3.2 million. By 2040, annual deaths are forecast to hit 4.1 million. You'd think the big casket makers—Batesville, York, and Aurora, which together produce 70 percent of all caskets sold in the United States—would be resting easy.

You'd be wrong.

In 2003, 76 percent of deaths involved the use of a casket, according to the Casket & Funeral Supply Association. That'll change. By 2025, nearly half of deaths will end in cremation instead of a full-body burial.* Of course, some families who wish to cremate may choose to first conduct a traditional viewing for their loved ones (as funeral directors strongly recommend), which might involve the purchase of a casket. Even then, few are likely to shell out thousands of dollars for a box destined for the incinerator.

To make matters worse, this shrinking market is also sought by a growing number of competitors in the form of discounters and independent craftsmen who ply their wares directly to consumers via storefronts or on the Web. Add to that the rising costs for steel and lumber, and you begin to hear the opening lines of an industry obituary.

Not so fast. For now, news of the casket's death is premature. With families still willing to pay $2,000 on average for a casket—a third of the average bill for the total funeral, which the National Funeral Directors Association says was $6,500 in 2004—there's still plenty of money in building boxes for burial.

The leading casket maker in the United States is Batesville. Founded in 1884, Batesville Coffin Company—so named for its hometown in Indiana—was bought by local businessman John Hillenbrand in 1906 (it's now part of Hillenbrand Industries, which also makes medical equipment). Now known simply as Batesville, it is the best-known, most respected mass-produced

---

* Cremation Association of North America.

brand in the business—the Toyota of caskets, if you will. Indeed, the company adopted manufacturing techniques used by the Japanese automaker in the early 1990s to improve quality, speed delivery, and reduce costs.

Today, company factories complete one casket every fifty-three seconds. Caskets are usually delivered within forty-eight hours. The company employs six hundred casket designs in 150 color combinations and thirty shapes. It owns more than 40 percent of the market share.

Batesville dominates the casket expo at the funeral convention. The display is less a booth than a stand-alone building, with elegant plywood interiors and a slowly revolving merry-go-round of caskets. It's so crowded, I can barely wedge the baby stroller inside. CEO Ken Camp races around wearing a Batesville polo shirt, shaking hands and handing out business cards. Time was, funeral directors would come to him to tour Batesville's factories and bond with its sales force. Times being tough, casket makers rely on these conventions to make an impression.

And Batesville has a plan to bring the mountain to Muhammad: its first ever traveling casket tour.

⬥

In May 2005, the Batesville trailer arrives in the parking lot of the Radisson in Saddle Brook, New Jersey. It has to be the biggest thing on eighteen wheels I've ever seen. The glossy black trailer is fifty-three feet long and has wings that expand its floor space to one thousand square feet. I've lived in smaller apartments.

The trailer's exterior carries the Batesville slogan: "Helping families honor the lives of those they love." Under the slogan are poster-size portraits of a U.S. Marine in full dress, a young family, an older couple, a fire fighter—the kinds of people who choose Batesville caskets, I presume. Five months into the tour, two thousand funeral directors in thirty-four cities have passed through this trailer or its twin winding its way up the West Coast.

Inside, ten local funeral directors have gathered for the 1:00 p.m. tour, led by a black-suited Batesville sales representative named Meghan Lopatich. The tour begins with a videotaped welcome from the CEO, shown on a flat-screen TV. Then we turn forty-five degrees to watch another slickly produced video on yet another flat-screen TV not four feet from the first (there is, around the corner, still a third TV). In this clip, a funeral director with Ken-doll hair expounds on Batesville's quality service. We see footage of nice-looking folks at a Batesville factory building caskets and answering phones. "How do we do it? Simple," says the voice-over. "It's our people."

Just as it would in a funeral home's casket showroom, the exhibit begins with the top-of-the-line model. About a quarter of a casket butts out from a wall (this being a traveling showroom, only partial caskets are displayed). As Lopatich steps toward it, a few of the funeral directors draw their breath. It's the Maserati of caskets: the Marsellus 700 Masterpiece. It's solid mahogany and 540 pounds empty. Ronald Reagan was buried in one, as were Mickey Mantle, Emily Post, and the Notorious B.I.G. It takes four weeks to complete the forty hand-rubbed finishes. "The light in here does not do it justice," Lopatich says with a sigh. Retail price: $18,500.

While the Marsellus represents the utmost in tradition (the 131-year-old line was acquired by Batesville in 2003), the rest of the trailer tour tries to show how Batesville is keeping up with modern funeral trends. Four trends in particular have rocked the American casket industry: obesity, personalization, cremation, and cost competition.

We start with obesity. According to the American Medical Association, it's linked to 300,000 deaths a year. If 76 percent of deaths involve caskets, you have to figure that's up to 228,000 oversize bodies needing boxes. Technically, all but the truly obese can be squeezed into a standard twenty-two-inch-across casket, according to funeral directors who've done it; one told me of shoehorning a beefy Philadelphia politician into a regular-size casket so that he could be buried in a historic cemetery with narrow plots. But a jammed-in corpse doesn't exactly make for an ideal memory picture at an open-casket funeral.

Then there are the truly obese. In May 2005, Goliath Casket Co. of Lynn, Indiana, made the news for building a casket to fit the body of a nine-hundred-pound man. It was seven feet three inches in length *and* width, used eleven yards of fabric for its interior lining, and required sixteen pallbearers to lift.

Batesville launched its politely named Dimensions line in 2004 at the Nashville convention (the display in the trailer is even more politely titled "Fitting the Times"). Although Batesville had built size XL caskets before, the ballooning demand had convinced executives to introduce more variations and bigger sizes.

The line ranges from the just slightly oversize models, twenty-three inches wide inside, all the way up to the truly supersized thirty-eight-inch version. Lopatich urges the funeral directors to carry the smallest model in their showrooms, both to save space and save the family from the humiliation of being trotted over to see a casket the size of a baby grand. She does not address how the family should cope on the day of the funeral, when the baby grand is on full display.

The next station in the showroom—titled "Infinite Possibilities"—tackles personalization. Companies like Batesville are at a disadvantage here; mass-produced caskets are by definition not easy to individualize. Its solution? Doodads. Batesville would probably prefer I use the brand name, LifeSymbols. A fishing enthusiast, say, can have a little fish ornament affixed to the front corner of his casket. These are called LifeSymbol Corners. A gardener could have a commemorative panel embroidered on the inside of the casket lid with a picture of potted flowers and spades.

I am a little put off by the doodads. They look tacked on, like an afterthought. Mostly I am bothered by the narrow selection. Can so many lives really be summed up by a handful of hobbies—golfing, sewing, or cooking? I think about my own dad, whose enthusiasm for tennis I suppose could be represented by this little racket and ball. But he's also a Catholic and an advertising man and an expatriate, not to mention a husband, father of four, and grandfather of eight. He loves hot sake and the Philadelphia Eagles and doody jokes.

How could a tennis racket alone possibly tell his whole story?

Come to think of it, he would probably most enjoy a LifeSymbol Corner shaped like a toilet, but I don't see one in the display. You'd think there'd be at least one doody doodad.

Embedded in the lid is a little drawer ten inches wide and about a foot deep in which one could place some of Dad's favorite things; the one on display has a golf ball and some tees. Lopatich says the MemorySafe

Drawer is quite popular—no surprise, as humans have been burying cherished possessions with their dead since before the Egyptians. Jews and Muslims aren't supposed to be buried with mementos, but Christians like to tuck Bibles or rosaries in the drawers, while secular Americans jam everything from cigars (for that first smoke in heaven) to cell phones (in case of boredom) to checks for $1 million (to go out a millionaire).

We move on. The next station deals with cremation. The company introduced its Options line of urns and other cremation products in 1993, when it became clear the trend wasn't going away. Its latest additions to the line even have some panache.

"These are inspired by Pier 1 and Crate & Barrel," says Lopatich, picking up an urn and handing it around. "They've got that bamboo, Tommy Bahama theme." Leaves have been decoupaged in layers on a wicker-style exterior.

"I call this my tissue-box urn," she says. "Seriously, I would use this as a tissue-box holder."

Urns alone won't cut it in today's cremation merch market. Batesville also partners with Nambé, a company Lopatich says specializes in wedding silver but also makes cremation jewelry and keepsakes—shiny pendants in the shape of hearts and teardrops designed to hold an ounce of cremains and be worn by a survivor.

Lopatich tells of a family that came into a funeral home asking for a direct cremation—a worst-case business scenario for a funeral director. But later the family ordered eight of these keepsake necklaces, and the funeral home wound up earning more than on a standard burial. There are hopeful smiles all around.

At the final station, Lopatich begins with a trick question. "Which one of these is made of wood?" she asks, standing before a display of nine casket butts. All, it turns out, are veneer. The use of veneer in casket making may seem unremarkable; after all, the cost of cherrywood has risen 80 percent in the past decade, as Lopatich points out, and veneer can cut costs dramatically. But the funeral business in America has never been about cutting costs. Biases still prevail.

"We hear veneer and we think, Ew," says Lopatich.

To my eye, the Glossy Maple and Pecan and Cherry and Oak look just like the real thing. Me, personally, I'd prefer to save the couple of hundred dollars that Batesville says veneer knocks off the price tag. Who would know? And who would dare to ask?

For all of Batesville's variety, though, what I don't see here is the option of simplicity. If I were to go the whole-body burial route, that's what I'd want: a plain cypress box, just like Pope John Paul II's. What a beautiful piece of work that was—the blond wood, the dovetailed corners, the simple cross affixed to the lid. (Before you judge me, I wasn't the only one who watched the papal funeral for the merch; soon afterward, funeral directors around the country reported requests for similar caskets.)

Today's consumers are used to demanding what they want. And various though their sizes and doodads and keepsakes and materials may be, a traditional casket maker's offerings still may not be enough to satisfy the tastes of a generation that went from hippie to yuppie, VW vans to minivans, Bush to Clinton to Bush. Ernie Wolfe, card-carrying member of said generation, offers an interesting alternative.

ERNIE WOLFE
1951–

*E*rnie Wolfe is late. The fifty-three-year-old art dealer was out diving for lobsters when I visit him this December morning, and his vintage truck has broken down on the way back to his Los Angeles gallery. So his wife, Diane, and I sit around a bowl of pistachios, waiting, surrounded by caskets and portraits of Arnold Schwarzenegger.

Wolfe sells African art. Caskets, it turns out, are the artistic province of Ghana, the tiny country on the west coast of Africa. Wolfe collects, imports, and sells them to Americans. (The Governator, it also turns out, is a beloved, near mythical figure to many Africans. Wolfe commissioned the portraits after the actor's California gubernatorial win, winding up with dozens of five-by-seven-foot canvases showing Schwarzenegger in various real and imagined poses—here at a podium with a manly Maria Shriver, here carrying a buck-naked woman in a nonexistent scene from the movie *Predator*.)

His most famous casket is one he's keeping for himself. When Wolfe finally ambles in, a big, sunny dude in shorts and flip-flops, he takes me straight to it. "Yep," he says after a moment of reverential silence. "This here's my personal ride to the great beyond."

We stand before a nine-foot lobster. It's propped up vertically in a dark corner of his gallery, painted bacteria green, protruding massive claws and skinny legs. Inside, says Wolfe with particular pride, there's fuchsia tuck-and-roll brocade lining. "Not that I'll care about comfort," he says. For one thing, he'll be dead. For another, he plans to be cremated, after which a can of his ashes will take up residence in the lobster.

Americans are only now discovering what Wolfe calls FAVs—fantastic afterlife vehicles. In Ghana, generations of artisans have produced caskets intricately carved and brightly painted in the design of something that represented the life of the deceased: a trowel for a mason, a sneaker for a soccer player, a bottle of beer for a bartender. Wolfe has collected caskets in the shapes of boats and tomatoes and airplanes and snakes. There's a whole subcategory of cars. Ghanaians are obsessed with autos, particularly Mercedeses. Wolfe even had an exhibit of car-shaped caskets in L.A.'s Petersen Automotive Museum until a janitor detected termite droppings.

Wolfe first stumbled upon FAVs in the mid-1970s while working as a scuba instructor and safari guide in Kenya. During his travels around the continent, he came upon a village in Ghana where wizened old artisans and their sons and grandsons fashioned colorful hand-made caskets. Over the years, he returned to purchase the best of these, becoming, he boasts, the first American to import Ghanaian caskets, selling them through his L.A.-based gallery.

One catch for the eager buyer: These caskets are not meant for the underground. "No way, man," says Wolfe, shaking his shaggy blond head. "I'm not going to that coffinoidal zone. My job is to protect them. If they go six feet under, what about us topsiders? How do we appreciate them?" In Wolfe's estimation, a wing of the Smithsonian Museum would make a more fitting final resting place. The museum, he says, has expressed interest.

There was one time Wolfe almost made an exception. An extremely wealthy customer had purchased a 1958 Corvette casket. It was only when Wolfe got the delivery order that he realized it was headed to Forest Lawn Cemetery. "That's when I found out he was dead as a doughnut," he says. "But then I thought, You know what, what the hell. At least it'll be admired at the funeral." However, the man's Catholic family balked, and the casket was returned. "I was pissed. I put a ton of work into fixing it up—replacing the nails and whatnot."

It wasn't the first time his caskets met rejection. A major museum backed out of an exhibit, sniffing that the work "smacked of idolatry." The memory gets Wolfe fired up. "Listen, idolatry—puh-lease. If you could see the love and care that goes into making them, and what they mean to the people . . . well."

In Ghana, the more elaborate the casket, the higher the status of the deceased. Wolfe shows me a photo album he keeps of Ghanaian funerals he has attended. For a fire chief's last ride, craftsmen had created a casket in the shape of a shiny fire engine, carried by a parade of goose-stepping, uniformed officers. When Sowah Kwei, the son of the man they call the father of Ghanaian caskets and himself a revered casket maker, died a few years ago, his friends called Wolfe to tell him of the arrangements.

"I said, 'What kind of casket are you making?' They said, 'Plain, plain.' I said, 'What do you mean, *plain*? Here's a revered artist, a master craftsman, and you've got a boring, plain casket?' Well, I got there, and it turned out they were saying *plane*." Photos show a seven-foot box shaped like a carpenter's plane, with saws and hammers carved out of wood and affixed to the sides.

In the Ghanaian funerals Wolfe has witnessed, the casket takes center stage. First the body lies in state at home for an extended wake. Then it is placed in the casket, which is taken up on the shoulders of young men, who dance through the streets, spinning it over their heads. Sheep and a bull are slaughtered. Guns are blasted, songs are sung. Bands of professional mourners, elderly ladies dressed in red, provide the sound track of wails. Finally, the procession winds up at the dunes, where the FAV is swallowed up by the sand.

The funeral of David Stein was less theatrical, but his ride no less magnificent. Stein was the father of Diane Wolfe, Ernie's wife. Diane says her father had stated a wish to be cremated and have his ashes scattered at sea. Her father's wife recollected differently, insisting he had wished to wind up alongside her in her already reserved crypt. They compromised. Diane had her husband order up an FAV in the shape of a black 1957 T-bird, about the size of a cooler. Diane and her siblings carried the T-bird containing his ashes up to the crypt, posing their kids alongside it. In the photos, the kids are grinning. The T-bird had an almost magical effect on the funeral party, Diane remembers, the kids in particular. "It made the whole thing seem more . . . I don't know, accessible and acceptable," she says.

So, too, might Ernie Wolfe's lobster casket. The Wolfes and their two sons—Ernest, thirteen, and Russell, eleven—live upstairs from the gallery in a loft apartment, and the boys use the gallery as their playroom.

"They've grown up around my casket. They say, 'Yeah, yeah, Dad's gonna be buried in that,' " says Wolfe. " 'And Grandpa's in a '57 T-bird.' To them, it's totally normal."

❖

*H*ere's the problem with buying a casket. Most people don't want to shop for one before they absolutely have to. And by the time they absolutely have to, they have a few days at most to make a major purchase of a product they know nothing about—under severe emotional distress, to boot. It would make so much more sense to have a casket at the ready. But there are practical concerns with that, too. How many people have the space to store a seven-foot, two-hundred-pound box? Where would you put it: the attic or the basement? What about termites, moths, or rust? Do they come with child-safety locks? Is it in bad taste to employ it in the meantime as a coffee table?

Then there are the psychological hurdles. Buying a casket is tantamount to admitting your mortality. Having it in the house is a daily reminder. If

58 percent of us can't even get ourselves to write a will,* how many of us can be expected to stock up on death merchandise?

My own feeling is that we'll remain a culture that makes death arrangements when and only when we must, including the purchase of a casket. Funeral directors and the major casket makers hope this means consumers will continue simply to pick one from among the models available through the funeral home. While I agree that most of us aren't yet ready to spend Sunday afternoon browsing casket stores, we are ready to window-shop—online.

Start with Google, where the search word *caskets* yields 1.9 million results. Batesville's own Web site leads the hit parade, but consumers can't buy directly from the company. For that you could visit direct-to-consumer outlets like FuneralDepot.com, which claims it will knock 70 percent off Batesville caskets. There's TheCasketCompany.com, which promises thousands of dollars in savings and free next-day delivery.

The Internet is an ideal tool for bargain hunters, and, to be sure, caskets can be had on the cheap online. But the Internet is also ideal for the selective shopper. TrappistCaskets.com sells the handiwork of the Trappist monks of New Melleray Abbey near Dubuque, Iowa. The monks sell about 120 pine or walnut caskets a month at prices ranging from $775 to $1,975, according to the Web site, using wood from the abbey's own forests.

Pretty as a plain pine box hewn by the Trappist monks may be, I still don't want one just taking up space in my bedroom. The solution: dual-purpose caskets. CasketFurniture.com is a Canadian business that builds caskets that double as sofas, entertainment centers, and pool tables. Down to Earth sells the wares of Bill Hale, a cabinetmaker in New Hampshire who also specializes in multipurpose caskets. "You can mothball your box at a funeral home," he suggests in a press release, "or turn it into a coffee table, gun rack, or bookcase—so you can use it more than once."

Even the KISS Koffin—a metal casket sold online that comes shrink-wrapped with pictures of the legendary hair band—has a here-and-now use. Take a guess. Duh: beer fridge.

Shrink-wrapping metal caskets in imagery meaningful to the deceased is a growing trend, if the selections online are any indication. A company called

---

* Survey, May 2004, Harris Interactive for Lawyers.com.

ArtCaskets.com offers caskets in NASCAR, military, hunting, and farming themes. One model, sheathed in an idyllic image of a golf course, is called Fairway to Heaven. Another is stamped with a message: Return to Sender.

Window-shopping online for caskets, however, is kind of like browsing magazine ads for pharmaceuticals: You can look all you want, but in the end you've got to go through a professional. Many casket makers, including Art-Caskets.com, sell only through its "network of distributors"—meaning you'll still have to go through a funeral home in most cases to purchase your Last Haul trucker casket.

Same deal for the cowboy caskets I saw on display at the convention in Nashville. I coveted one called Santa Fe, a rectangular box made of red cedar, its lid inlaid with nuggets of turquoise; I could see it pulling double duty as a blanket chest. No go: The representative at the convention tells me its products are offered "exclusively" through funeral homes. The Web site, Cowboys-LastRide.com, lists dozens and dozens—in Texas. Says the company in a statement: "Funeral homes are a godsend and provide very special services. We are not in the business of providing or competing with these services."

———

The possibility that consumers might want to buy caskets on their own is a matter of grave concern to the funeral industry. That's why what happened in August 2004 shook the business like a bomb: Costco began to sell caskets.

On its Web site and at a handful of stores in Illinois, the discount retailer markets coffins directly to consumers. At Costco.com, caskets are easy to find; it's the shopping category tucked alphabetically between "Books & DVDs" and "Computers & Peripherals." When you click on the link, up pop six models made of eighteen-gauge steel in colors like lilac and Neapolitan blue. All are crafted by a company called Universal Casket Co., delivered within three business days, and cost $799.99 including shipping.

The reaction from the funeral business was swift and shrill. The Illinois Funeral Directors Association issued a press release in September 2004 titled "Bargain Caskets May Not Be Such a Deal." It warned of hidden costs, substandard quality, and the dangers of purchasing big-ticket items online in a time of emotional fragility. Paul Dixon, the association's director, had this quote about casket discounters: "I've heard of many cases where a family has thought that they were saving money, only to find they could have saved more money and obtained a better quality casket by purchasing the casket through their local funeral home."

A survey by FuneralWire.com, an industry Web site, showed 52 percent of funeral directors thought that "Costco and other large retailers should be allowed to sell funeral merchandise," while 45 percent felt they should not.

The phrasing of that question struck me as odd: Should Costco be *allowed* to sell caskets and other merchandise? Who died and made funeral directors the God of all retail? But then again, this is the death industry we're talking about, where the normal rules of commerce have long failed to apply—and the sale of caskets is perhaps the best example.

Here's how it works. Casket makers persuade funeral homes to push their wares, dangling discounts for exclusivity. Samples are displayed in the funeral home's showroom and catalogs, to be seen by families arriving to make arrangements. Typically, families choose from the selection on the floor; that casket is then replaced by the casket manufacturer. The family pays the funeral home at the retail price, out of which the funeral home pays the casket maker the wholesale price.

So far, that's not so different than buying a Toyota from an auto dealership. If the dealer makes a profit off the car, so be it. But what if that Toyota dealer was the only place you could buy your wheels, and it could refuse to let you drive a vehicle you found elsewhere? Worse—what if the markup was seven times the wholesale cost?

That's what consumer advocates charge has long been the accepted business practice of casket makers and funeral homes. "The whole idea is to keep consumers in the dark about their options," says Josh Slocum, executive director of the Funeral Consumers Alliance. "By scaring them out of comparison shopping, they're effectively killing fair market competition." The Federal Trade Commission's 1984 Funeral Rule insists funeral homes must allow a family to use a casket purchased elsewhere and may not charge a handling fee. But in his reviews of funeral homes' general price lists, Slocum says he frequently uncovers just such a charge.

Costco isn't the only one presenting a looming price threat for major casket makers. As with our diapers and toys and pullovers, China is proving adept at turning out good-looking, low-cost caskets, priced at half or even a third the cost of U.S. models. At the Nashville convention, there is a booth of Chinese caskets with a display model that looks exactly like one I see later over at Batesville.

As of now, Chinese imports make up just 2 percent to 3 percent of the market, according to the Casket & Funeral Supply Association. And there are disadvantages for the importer to consider: The distance makes prompt delivery out of the question, and their factories are not yet equipped to handle customization. Still, any consumer will tell you that a "50% off" label on the price tag is a pretty attractive lure.

Teon Austin knows this. Austin is a muscular African American man with a shaved head and a calm manner. By day, he's the proprietor of a small store in Clifton, New Jersey, called Competitive Caskets. By evening, he's a store detective for Home Depot. By night, he's a bounty hunter. Teon Austin is a reality show waiting to be pitched.

The casket store resides in a small shopping square called Botany Village, in between furniture and video stores. There's also a supermarket, a Dunkin' Donuts, and a Subway in the square. That means a regular stream of shoppers pass a window display crowded with urns. Just behind the plate-glass windows, in plain sight, is a row of open caskets. The shop does not try to disguise its purpose. Its elegant green awning reads:

COMPETITIVE CASKETS
50%–70% OFF
Funeral Homes Must Accept Our Caskets

Austin sits at a lone desk just inside, tapping at a Toshiba laptop and talking on his cell phone. The showroom is a long, narrow, one thousand-square-foot space, with fluorescent lighting and drab carpeting. Eight caskets line one side of the room.

There's the Francis, a sixteen-gauge "sealer" with a velvet interior. "Funeral Homes: $6,600," reads the sign. "Our price: $2,599." The Katie Pink, twenty-gauge with crepe lining, is $2,400 in funeral homes, $1,160 here. The cheapest casket is $699; the most expensive, $6,716 (though Austin has yet to sell one of those).

"What I'm here for is to save the public some money," he says, joining me by the caskets. "We are not a company that rips people off. I'm not saying funeral homes rip people off, but they do charge an enormous amount for caskets. You can see that I don't. Some folks just don't have that money. I give them the same casket, same quality—for a whole lot less."

Austin says he got his first exposure to the business of casket sales as an employee at funeral homes. When a friend opened Competitive Caskets in 1999, he came on board as a salesman, then bought the business outright in 2000. The store has caught its share of grief from local funeral homes.

"I have a good rapport with some of them," says Austin. "But others—I won't say who—they tell their clients not to come here. They say things like the bottoms will fall out of my caskets, or I'm shady. Like that."

Sometimes the resentment is palpable. When he carts the heavy caskets into funeral homes, "some of them just look at you like they want you to drop it and damage it." Austin asks the funeral directors to sign forms as proof the casket arrived in good shape.

But far from harboring ill will toward the funeral business, Austin hopes to go to mortuary school and own his own funeral home within the decade.

"I'm not making a killing here," he says.

Last year, he sold 115 caskets at an average of $1,500 to $1,600. That means Competitive Caskets grossed about $178,000 before wholesale payments to the casket maker, an independent craftsman in Pennsylvania.

It's not that Austin won't sell the major brands. He'd like to. But every time he calls to inquire about carrying their wares, when they learn he's an independent retailer they hang up.

On the face of it, this makes no sense. Why won't casket makers let legitimate outlets like Competitive Caskets or Costco sell their products?

Wouldn't it improve their sales? I ask Batesville spokesman Joe Weigel why the company sells exclusively through licensed funeral homes.

"It's something we have been doing for many, many years," he says. "From a business standpoint, if the business model is working, why change it?"

Funeral directors make better salespeople, he adds, because "they go the extra step of staying up on caskets. They take the time to go to tour centers to learn. They are there on a permanent basis. If a family has a problem with a casket, they know that if they purchased from a funeral home, the funeral home provides that permanence."

I point out that funeral homes can go out of business like any retail store.

"Well," he says, "they have very, very strong permanence."

Some suspect darker motives for keeping the business model. A lawsuit filed in May 2005 by the Funeral Consumers Alliance charges that the three largest funeral home chains—Service Corporation International, Alderwoods Group, and Stewart Enterprises—were fixing prices on Batesville caskets, as well as urging families to buy their higher-priced products by conducting a smear campaign against casket discounters. The lawsuit, which had yet to go to court in late 2005, also names Batesville for agreeing not to let discounters sell their products in exchange for exclusive access to funeral chains' clientele.

"A lot of people in the industry are saying this is all stuff that's been going on forever," says Josh Slocum of the FCA. "Well, then the question is, Why didn't someone do this before?"

Batesville will not comment on pending litigation, says Weigel.

The lawsuit, if won, may succeed in finally disassembling the casket industry's business model. I think it'll take something else: the choosy, cheap, loudmouthed American consumer.

In this age of on-demand everything, I don't see why I can't buy what I want from whom I want and use it however I want. How come it's harder to buy a casket than it is the gun that might put me in it? If enough of us demand that turquoise-studded casket for use in our living rooms and later at our wakes, then eventually the industry will have to give in. Capitalism works better than regulation anyway . . . isn't that what our political leaders tell us?

Now if only it would come with child-safety locks.

Disney on Ice

A TOWN CELEBRATES ITS FROZEN DEAD GUY

*Miniature skeleton people join in the wholesome family fun of the
Frozen Dead Guy Days festival in Nederland, Colorado.*

BREDO MORSTOEL
1900–1989 (AND BEYOND)

*I*n a town called Nederland, Colorado, outside a nuclear-bomb-proof house, inside a Tuff Shed, at the bottom of a large freezer, next to a half-eaten birthday cake, lies the body of Bredo Morstoel.

Or so I'd heard.

The story had been told and told again. Bredo Morstoel had died at age eighty-nine in his native Norway. His daughter Aud and his grandson Trygve Bauge had flown his body to Colorado, where the two were then residing. But Trygve, a budding entrepreneur in the field of cryonics, had decided to keep Grandpa around. Grandpa Bredo died in 1989, and in 2005 his body is still on ice in a Tuff Shed in Nederland.

When I first heard about Grandpa Bredo, I thought I would have to see for myself this curious experiment in human preservation. I thought Grandpa might teach me something about cryonics—how it's done, why we bother, what it says about us. But, as a townsperson says to me later, the preservation of Grandpa is the "1968 VW Bug of cryonics"—hardly worthy of the scientific category, even such as it is. What I stumbled into instead was a curious experiment in death celebration, or the story of a community that initially recoiled from death but came to embrace, laugh at, and profit from it. What I stumbled into was the Frozen Dead Guy Days festival.

Nederland is a no-stoplight town twenty miles west of Boulder, 8,233 feet above sea level. One spaghetti of a road takes you there, twisting and turning as it slithers up the snow-capped mountain. I am the white-knuckled, baby-on-board driver creeping up that road in an economy-size rental car as four-wheel-drive SUVs pile up behind me. The town has one inn and one supermarket and 1,380 residents—a disproportionate lot of them, shall we say, different. There's Toasty Post and the Iceman, and then there's Amy the ghost buster (more on her later). In a recent mayoral race, the field of candidates included a convicted felon and a dog.

It takes serious eccentricity to stand out in that company, as Trygve Bauge did—or, more important, it takes PR. Trygve was already known in the Boulder area for his self-publicized exploits running from immigration authorities (to avoid deportation for overstaying his visa) and staging a mock hijacking prank at the airport (to—oh, who knows why). The skinny, long-haired Norwegian eventually drifted up to Nederland to build his dream home: a concrete-and-metal bunker that would withstand nuclear, biological, and alien attack. No one can remember Trygve ever holding a job. He seemed too busy pursuing his two passions: ice-bathing, in which he claims to hold the world record at one hour, five minutes and fifty-one seconds, and the practice of what he calls "life extension."

Which brings us back to Grandpa Bredo. Not a whole lot is known about Bredo Morstoel, at least to the people of Nederland, being that he took up residence here only after his death. In the one photograph I've seen, he has tufty white eyebrows and ruddy skin, and he's squinting at the camera. He worked for the parks department in Norway, was married, and had two children. He liked to paint, fish, ski, and hike. During a family vacation in the mountains in 1989, Bredo had a heart attack while napping and died.

I learn much of this through the sleuthing of Barbara Lawlor. Lawlor is Nederland's Lois Lane, a one-woman media machine whose byline accompanies virtually every article and photograph in the weekly *Mountain-Ear*. Lawlor is deeply tanned and white haired and looks as if she could hike to Boulder and back in under an hour. A Wisconsin native who took the reporting job to support her four adopted children, she's not one to suffer fools—but a story is a story, so she wearily took Trygve's incessant calls. It was Lawlor he called to witness his record-breaking polar-bear dip in an ice-filled cistern wearing a pair of boxers and a Norwegian flag. It was at Lawlor's door that Trygve turned up one night demanding sanctuary from immigration officials. And, soon after Trygve was finally deported in 1994, it was Lawlor his mother, Aud, called with an odd and disturbing request.

"I remember it was Mother's Day," says Lawlor. "Aud came to me, crying, saying something about going down to Town Hall and getting help putting dry ice on her father." After some questioning, Aud explained the situation. Lawlor knew this was more than just a scoop. "Well, I went to Town Hall for her and said, 'There's a dead body on this property,' and they went nuts, and that's how it all started."

Police and town officials raced up the dirt roads to Trygve's bunker. Sure enough, in a Tuff Shed a few feet from the house was a wooden sarcophagus containing a lot of ice and the body of Bredo Morstoel. The shed also held the intact remains of one Al Campbell of Chicago, apparently the first paying customer of Trygve Bauge's backyard cryonics lab.

All hell, as they say, broke loose. The dailies and the wires beat Lawlor's weekly to the story, and within days Norwegian camera crews were camped out on Lawlor's floor. COLORADO TOWN FINDS 2 BODIES ON ICE; MAN HOPED TO REVIVE THEM, read a May 12, 1994, *Chicago Tribune* headline. The Associated Press followed up with FROZEN BODIES GET CHILLY RECEPTION FROM TOWN LEADERS.

Chilly? Nederlanders were horrified that their bucolic mountain town was now known in England and Japan for its resident frozen dead guy. In an emergency session of the Nederland Town Council, officials slapped together an ordinance heretofore outlawing the storage of dead human or animal parts on residential property—thereby effectively banning pork chops from freezers. The law didn't affect Grandpa, who, as locals like to say for the yuks, was grandfathered in.

So Grandpa Bredo remained in Nederland, outside a nuclear-bomb-proof house, inside a Tuff Shed, at the bottom of a large freezer, next to a half-eaten birthday cake. (Campbell's body was immediately shipped back to Chicago.) The Tuff Shed is new and improved, donated by the company, which belatedly but enthusiastically embraced the marketing opportunity. The rickety wooden sarcophagus has been replaced by a stainless-steel box. The bombproof house is currently uninhabited, but a local hired by Trygve visits Grandpa every five weeks to replenish his bed with eight hundred pounds of dry ice.

Or so I'd heard.

✣

It's a brilliant March day in Nederland, so warm that I don't need a jacket. I strap Mika to my back and head to the visitors' center. Ressa Lively-Smith, a tall woman with flouncy gray hair and a cowl-neck sweater, is manning the booth. There is a steady stream of visitors to the center, and everyone wants to know about one thing: the frozen dead guy.

This is the start of the Frozen Dead Guy Days festival, Nederland's biggest tourist attraction of the year. There's the Cryogenic Parade down First Street, the Frozen Dead Guy Lookalike Contest, the Costumed Polar-Bear Plunge, and of course the Coffin Race (brought to you by Tuff Shed).

Frozen Dead Guy Days irks Ressa Lively-Smith. A thirty-six-year resident of Nederland, she was one of those who stood up at the town meeting professing outrage at the body's presence. "'Course, I'd had a few Jack Daniel's by then," she says. "I just did not think putting someone on ice in a Tuff Shed was a hygienic or kind or loving thing to do."

The phone in the visitors' center rings.

"Well, you know this is the weekend they do the festival for the frozen dead guy," she tells the caller. "Yep. That's right. Someone froze a guy and put him in a Tuff Shed. Let's see, there's a Blue Ball—because he's a he, get it? And people who are really, really on the brink jump into a pond at the edge of town. . . . Yep. I know. Well, Frozen Dead Guy Days is happening whether you or I like it or not."

She hangs up and shakes her head. "Many people say the festival is at the very least odd, at the most morbid and grotesque," she says to me. "But hey, it's March. We have to get the people up here. And oh, I tell you, they come from all over. Lots of Norwegians, of course. And Orientals and Europeans."

Getting bodies—live bodies—up to Nederland in March is precisely the point of the Frozen Dead Guy Days festival. Teresa Warren, owner of Off Her Rocker Antiques, was president of the Nederland Chamber of Commerce in 2002, the year the festival began. I meet her at Grandpa's Blue Ball, held at the Black Forest chalet restaurant, at which the Ice Queen and Frozen Dead Guy Lookalike will be chosen.

Warren is dressed in her Ice Queen getup of purple-tinsel wig, black lipstick, and strapless blue ball gown. Her husband, Brent Warren, is dressed as a cryogeneticist gone wrong, cradling a puppet of a rotting old man swaddled in a baby blanket.

Teresa Warren is telling her version of how the festival began. "We were talking about doing a winter carnival to attract tourists and calling it March

Madness," she recalls. "I told my husband about it and he said, 'You've got this great gimmick. We're known for Grandpa. Make it Frozen Dead Guy Days.' " Perhaps it wouldn't be so strange, they reasoned; after all, there was the Mike the Headless Chicken Festival in Fruita, Colorado, in honor of a chicken that lived for over a year after its head was cut off in 1945. When she brought the idea to the chamber, it turned out a few other people had had the same epiphany. "We decided the mystique wasn't going to go away, so we might as well make a few bucks on it."

Understand that this was a seismic shift for a town that ordered Grandpa's body removed within thirty days after its discovery. The mayor at the time had shrieked at the town board meeting, "I won't have this made into a dog-and-pony show!" But the years and the exposure had mellowed the townspeople. B. J. Ball, the only born-and-bred Nederlander I meet, says he balked initially at the suggestion of a dead guy festival. "I thought surely there's something else we could be known for," says Ball, whose family co-owns the town grocery store. "But I figure it's better to laugh at yourself. It's only as morbid as you're going to make it. After all, what's Presidents' Day if not a day to celebrate some dead guys?"

Maybe all that this proves is that people will embrace any unflattering image as long as it makes them money. The festival quickly became the town's biggest moneymaking week of the year. The Chamber of Commerce's annual budget went from $7,000 before the festival to $125,000 in 2005. Business quadruples for many local merchants, including Warren's antique shop. The Nederland Best Western posts no vacancies. The Mountain People's Co-op sells out of Grandpa's Favorite Muffins, and the urn labeled "Ashes of Problem Customers" at the corner coffee shop spills over with tips.

"It's been great for me," says Brent Warren, who is a graphic designer, filmmaker, event producer, and clown (Nederlanders wear many hats). He designed the program cover this year as well as the book jacket to a mystery novel set here in Nederland called *One Too Many Dead Guys*. He also produced the festival's opening ceremony, modeled after the Olympics, featuring a torch run from the Tuff Shed to the town center carrying "Grandpa's spirit."

The Black Forest chalet is getting crowded. Nearly everyone is in costume. The blind guy who won last year's Grandpa look-alike contest is over by the band, in a wig with his face painted white. Everyone knows he'll lose this year to a Belgian celebrity named Herbert Flack, who has the unfair advantage of a professional makeup artist to apply fake icicles to his nose. Flack is in town filming a popular Belgian TV show called *The Best Belgian,* in which the producers find the wackiest contests around the world and enter a Belgian celebrity. As a producer explains to the *Mountain-Ear:* "We bring the Belgian, and he is always the best Belgian because he is the only Belgian."

Sitting quietly at a round booth table, attracting little attention, is Aud Morstoel. This is Aud's first time attending the festival held in the name of her father. In fact, it is her first time back on U.S. soil in over a decade; she was deported not long after Trygve. Aud was granted a special visa to attend the festival this year, though her son, who also applied, was not; in the Norwegian press, he blamed President Bush. Trygve Bauge, sequestered in Norway and busy building his cryonics empire, had failed to respond to my e-mails requesting an interview. But here was his mother, live and in person.

Aud Morstoel is a woman of about seventy. She wears black, and her shoulder-length hair is windblown. She sits with her back slightly hunched and her hands around an almost empty glass of beer. I introduce myself and ask what she thinks of the festival.

"I think it's the right thing to do to get people interested in this dead guy in the coffin," she says, referring to her father. Her English is dusty and halting. "As long as it is friendly." I wonder if she really views events like the Frozen Dead Guy's Head Roll—in which frozen turkeys will be bowled down a street—as friendly. I guess she means good-natured.

The attention, she hopes, will spur scientists along in their life extension research. "They need to work so we can live longer," she says. "And also so we go to the stars."

The stars? I had not been aware of an interplanetary aspect to cryonic research.

Aud leans in and looks grim. "I think life on earth is over in a few years. My father, if he had lived today, would have wanted to go to the stars."

Did he say so?

"No," she admits. In fact, Grandpa hadn't mentioned any wish to be a

Popsicle, either. Trygve claims that Grandpa arranged his deathbed pillows in the shape of a T, indicating he wanted Trygve to arrange his care. "Trygve, he wanted his grandfather to live longer," Aud says sadly. "If he was younger, they said they would have given him a new heart."

I ask what she and her son plan to do with Grandpa. I am presuming they will not keep him iced indefinitely. I had read that the pair planned originally to send Grandpa's corpse to a cryonics lab in California. At ventures like Alcor Life Extension Foundation, perhaps the leading name in the field, bodies are immersed in liquid nitrogen, frozen to minus 320 degrees Fahrenheit, and entombed in ten-foot-high steel tanks. Alcor is home to the body of baseball great Ted Williams, whose family feuded bitterly over his interment and whose body was "outed" by a former Alcor worker in 2002 as having been decapitated. (Alcor has said that removing the head is standard procedure.)*

"I think we have to clone him," says Aud.

Back in Norway, Trygve is apparently continuing that pursuit. His delirious Web site, Trygve's Meta Portal, lists, among dozens of enterprises, each one followed with a ™ symbol, International Life-Extension Centers, the Rocky Mountain Life-Extension Center, and the Rocky Mountain Cryonics Facility. Aud appears to have rock solid faith in his—and science's—eventual success. She and Trygve intend to be "frozen the same way as my father."

By now Aud's speech is slowing, and I fear she is going to fall asleep. Then she says something kind of poignant. "My son has difficulty, but he will never give up," she says. "He had a lot of trouble to get my father to the U.S. So it is like the, what, the Coffin Race. It has many . . . many . . ."

Obstacles?

"Yes, many ob-sta-cles and has many troubles to win. It is same for Trygve and my father. There are many ob-sta-cles, but we will win."

As I get up, I ask if I may buy her a drink. She points at the empty bottle next to her glass and, for the first time, smiles. She's drinking Dead Guy Ale.

---

* "Please Don't Call the Customers Dead," *New York Times*, February 13, 2005.

It's nearly 10:00 p.m., so I leave the Black Forest and walk to the B&F supermarket. There, a dozen or so people, all young, mostly coupled, have gathered for what is billed the Moonlight Champagne Frozen Dead Guy Tour. This tour costs $35, ten bucks more than the regular daytime tours, presumably for the bubbly and the spooky nighttime atmosphere. I suppose it would make for a memorable date.

The nervous chatter dies as Bo Shaffer comes barreling in through the sliding doors. "Okay," he says, clapping his hands. "Who's ready to meet the dead guy?" Shaffer wears a mustache, a cap that reads, "Dead Guy," a black leather jacket, and a Stars and Stripes scarf. "Do we got the champagne?" The van driver nods.

I ride up ahead with Shaffer in his red pickup truck. It's plastered with bumper stickers that say things like "It's the Second Amendment That Makes All the Others Possible" and "Fight Crime—Shoot Back."

Shaffer is the dead guy's keeper, the one they call the Iceman. Every five weeks—four in the summer—he trundles up the treacherously steep, unpaved roads in his pickup to deliver the dry ice. The ice has kept Grandpa Bredo's corpse at negative 109 degrees, more or less, since his death. (I say "more or less" because Grandpa has thawed out at least once, when the poorly constructed original coffin sprang a leak.)

On the way there, Shaffer tells me he has a background in science and ecology and that he conducts home inspections for a living. I ask if he had ever kept a dead body before.

"Well, no," he answers. "But what we're doing here is maintaining an artificial environment, and that, I've done. Does it take high tech? Not necessarily. Sure, it's nice to have stainless steel and liquid nitrogen. But you can make do with wood and Styrofoam. It's backyard cryonics, but it's perfectly viable." Shaffer talks fast and loud like a carnival barker; I smell a little alcohol, but maybe it's spilled champagne.

It is pitch black up at Trygve Bauge's house. If you can call it a house. It looks more like an unfinished bomb shelter, and in fact that's what it is: Trygve and his mom claim that, when completed, the cement-and-steel structure will withstand the nuclear war they insist is imminent. Shaffer leads the way inside, where he lights a gas lamp. There's no electricity. Still, "It's earthquake-proof, fireproof, atom-bomb-proof—everything but waterproof," says Shaffer, looking up at the stars through a gap in the ceiling.

He herds us around a table constructed of planks that acts as a souvenir shop. There are pieces of the original Tuff Shed, the one they tore down when the company donated a spiffier model. The pieces of jagged metal sell for $20, complete with a certificate attesting to their authenticity. If you claimed to own something like that, would anyone ever ask for proof?

For a few bucks more, you get to have your name inscribed in Grandpa's current shed ("so that when he wakes up, he'll know who to thank," says Shaffer). There's a CD of death-related songs called *Dead and Dying*, which plays from the boom box; stickers that read, "I ♥ Cryonics"; plastic toy coffins full of candy; photos of Bredo, $2 to $5 depending on size; and Dead Guy Ale hats ($20) that "can be autographed if you wish."

I look around. Autographed by whom?

Shaffer doesn't miss a beat. "If you turn your back, we can probably get Grandpa to autograph them," he says. "Otherwise, I would, I guess." He hands out Frozen Dead Guy press-on tattoos. "You know, for the kids."

Outside, the March wind is whipping, and the couples huddle together. Shaffer points out the giant cement tubes that were to make up the tunnel Trygve meant to build from the house to the Tuff Shed. Nuclear protection for Grandpa, it seems, was not a top priority. The shed itself—Grandpa's home—stands a good fifty feet from the house. It's covered with ads from sponsors like Penguin brand dry ice and the local chapter of Libertarians.

"You can see we're nothing if not marketing geniuses," says Shaffer. I can't tell if he's being sarcastic.

Slowly he opens the double doors. Most of the space inside the shed is taken up by a rectangular wooden container. It stretches end to end about ten feet and stands about chest high. Shaffer lights a flickering flashlight. We all squeeze inside, two deep surrounding the container.

The doors creak closed, and someone squeals. This pleases Shaffer. "I purposely don't oil those hinges," he says.

Enlisting the help of some of the young men, Shaffer heaves open the container top. "Everybody else, move back!" he hollers. "I can't be responsible for what jumps out." A girl clings to her boyfriend. Smoke from the dry ice puffs out. Someone coughs.

After a moment, we all peer inside.

It's an icebox, a giant, primitive fridge on its back. Underneath the smoke there are chunks of dry ice, then more dry ice. There's a pillow and a comforter, both frosted white. And what appears to be a frozen cake.

We glance quizzically at Shaffer, who begins to laugh. It's a full-bellied, I-was-waiting-for-this guffaw. I get the sense he is laughing at us.

"No, you don't get to see the body," he says, clucking. "Everybody asks if you get to see the body, and no, you just see the sarcophagus."

I feel it is a little late into the tour to impart this information. Some in the shed are clearly disappointed. Shaffer moves aside the comforter to reveal a metal box wrapped with a thick chain. It is a stainless-steel, hermetically sealed sarcophagus, he says. Grandpa lies inside. This is as close as we will get to him.

Shaffer pops the champagne while telling a story about how, on the inaugural Moonlight Champagne Frozen Dead Guy Tour, he naively kept the bottle on dry ice, only to have the cork burst off and ricochet around the shed into a framed picture of Bredo hanging over the coffin, which fell onto a Tuff Shed executive, whose hand bled all over the floor. "I'm not sure why Grandpa would jump out and smack the hand of the guy who built him his house," he says.

He reaches into Grandpa's freezer and grabs chunks of dry ice, plunking them into the plastic glasses. I refuse. I could have used the alcohol, but champagne laced with ice used to keep a human body frozen . . . well, that's just gross.

"It's a unique tour," Shaffer concludes, grinning proudly. "We're only one of five or six cryonics labs in the world. Yes, we get a lot of grief over the fact that we're the 1968 VW Bug of cryonics, while you got the 2005 Cadillac Escalade over at places like Alcor."

We all stare down at the sarcophagus one last time. It seems small.

"He was fairly short," says Shaffer.

On the way back to town, Shaffer drops his circus ringmaster act. It turns out he's not at all the pom-pom-waving cryonics cheerleader I'd believed him to be. He's a spiritual wanderer and diehard Libertarian and workingman who makes a few extra bucks a month for his family by filling a box in a shed on a mountain with dry ice.

"It's the mortal coil, man," he says, upper lip curling in exasperation.

"This bit of meat, this piece of flesh—when you're gone, you're done with it. It's a worn-out space suit. I mean, get with the program, Trygve."

————

The program begins the next day with the Cryogenic Parade. Hundreds of people line First Street in front of the Rustic Moose gift shop, the Mountain People's Co-op, and the Pioneer Inn saloon. A lady in a sweeping black dress sells bouquets of dried flowers, shouting, "Dead flowers for sale!" This is a popular merchandising tactic around town: What's lifeless is simply rebranded as dead.

Aud Morstoel, the grand marshal, opens the parade riding in an open convertible. She squints at the crowd and waves. Participants in the Coffin Race prance by with their creations. Children dressed in skeleton suits throw candy. The hearse club of Greater Denver follows, showing off their lovingly restored, antique Cadillac deathmobiles. Their drivers have dressed for the occasion in bloody gauze bandages and ghostly makeup. The license plate on a particularly magnificent purple Oldsmobile hearse reads 1FOOTIN.

After the Costumed Polar-Bear Plunge, held in honor of Trygve in the frozen pond at the edge of town, there's the highlight of the festival: the Tuff Shed Coffin Race. Up to six people carry a homemade open coffin with someone sitting inside over an obstacle course consisting mostly of big piles of snow. The fastest team wins $300. One team has built a coffin gussied up like a tank; another team is dressed as bloodied doctors, their mangled patient sitting in the coffin; another inexplicably wears tutus.

A little girl and her mother watch a team practice mounting and dismounting their coffin. "Do you think someone died?" the girl asks.

Her mother starts to laugh. "Fair question," she says.

There are many families here at this festival of death, including those with small children. Most, when asked why they came, respond with a shrug that it just "seemed like a fun thing to do." I have met a college student from Front Range and middle-aged ladies from Ohio and a tattooed air traffic controller who rode here with his Harley buddies. At the Coffin Race, a grandmotherly woman in a blue sweater leans over to peek at Mika, who is falling asleep in her back carrier. Mary Wallach, fifty-seven, is visiting from Los Angeles.

I ask why she is here.

"We heard about it and said, Well, of course, we have to go," says Wallach. The festival is not an entirely unusual event on her itinerary; she and her partner, Diane Donile, fifty-nine, spend many a vacation visiting cemeteries.

Wallach motions Donile over. "What was that one cemetery we went to, that really beautiful one . . ."

"Pueblo? Deadwood? Key West?"

"Oh, Key West. That was a great one. You've got to see it," says Wallach. She pauses. "Don't get us wrong. We're not morbid or maudlin. Death used to make me fearful. Not at all anymore. It seems so . . . untragic. So right."

The two have their afterlife planned out: They will be buried in a plot next to Donile's husband, who died when he was forty-nine. "That was my first experience with cemeteries," says Donile. "I had no idea of his wishes. It was kind of like buying real estate. The lady talked me into buying the plot next to him. I was forty-seven."

"And now, if we want to be buried together, I have to sleep for eternity next to a guy I never knew," Wallach grumbles.

They will be cremated, they have decided. "I never wanted to be interred," says Wallach. "That old children's rhyme about worms crawl in, worms crawl out—ugh. That just never appealed to me." Not for her a scattering at sea, either. "Have you ever attended one? It's cold out there, and dark, and wet."

I say good-bye and walk over to a small, dark hall under the pizza shop to check out the Dead Guy Expo. It is not much. There are two bored-looking palm readers waiting for customers at foldout tables. Five people are in the back watching *Grandpa's Still in the Tuff Shed*, a documentary about the whole saga; the filmmakers, the three Beeck sisters from Boulder, hold court nearby. The busiest table sells merchandise including Dead Guy Days posters, T-shirts, and hats. For my husband I buy Dead Guy coffee beans from DazBog, a local coffeemaker, and for myself a little plastic ice cube with Grandpa's squinting visage lodged inside.

For the first time this year, the townspeople have told me with some excitement, there will be a real live cryogeneticist here. She turns out to be an African American woman named Kennita Watson who calls herself a "cryonics advocate." She leaps to her feet when I approach her table. "I'm here to declare that cryonics is a true end-of-life option," she says. She speaks and reshuffles her stacks of pamphlets with great energy, yet never quite meets my eyes. "I want to live a long time. Listen, a hundred years ago, there were no quintuple bypasses. Imagine what we'll be able to cure in another hundred."

She contacted Nederland on her own, she says, to travel here during the festival and clear up some misconceptions. "Bredo—now, that's not cryonics, actually," she confides. "He's just on ice."

<hr />

"What happens to our souls when we die?"

That's the opening line of *Grandpa's Still in the Tuff Shed*, the Beeck sisters' feature-length documentary. The original, short version had the backing of Michael Moore and made the rounds of independent film festivals. Initially banned by embarrassed Nederland town officials, the film won three awards and was eventually shown at school auditoriums.

A portion of the documentary is devoted to Grandpa sightings, with Nederlanders recounting ghostly appearances as spooky music plays in the background. I ask Robin Beeck, the film's director, if she thinks keeping a body around also keeps its soul around. In other words, by keeping Grandpa's body around, have Trygve and Nederland trapped his spirit, too?

In response, Beeck asks me, "Do you believe in ghosts?" She recounts a couple of odd things that happened to her and her sisters as they made the film—TVs suddenly flashing on, doors swinging ajar. But on the subject of Grandpa Bredo's ghost, she says, "You really have to ask Amy Bayless."

Amy Bayless is Nederland's resident ghost buster and the only person in town who claims to have had direct contact with Bredo Morstoel in the afterlife. I didn't need to go looking for her. Somehow—either through ESP or town crier Barbara Lawlor—she learned of my presence at the festival and tracked me down.

Often, during the course of reporting this book, the kooks I had been expecting to meet turn out to be deeply thoughtful, sensitive people who have much to teach me about death. Amy Bayless is such a person. Originally from Indiana, Bayless knows death intimately. It'll be sixteen years in June 2005 since she lost her only daughter. Denise, an army nurse, was twenty-three when she died in a motorcycle accident in Texas. Denise's father, from whom Bayless was long divorced, insisted on a military funeral. But Bayless needed to grieve in her own way. She took Denise's ashes to a lake in southern Indiana where the two used to go camping.

"That's where she is," says Bayless. "Her fiancé and I sat on the side of the hill. We sat around and told stories about her and drank tequilla till we were stupid. In our own way, we had our own party for her."

If it's possible for anyone to be equipped to handle such a devastating loss, Bayless says she was. "I hit a point in 1978, '79, when I looked around me and thought the American way of death was just screwed," she says. "When babies come into this world, it's such a sacred thing. As much as people benefit from a midwife when they're coming into the world, they need the same going out." She trained as a death midwife, volunteering in hospices to help ease the passage of dying patients. "When you die that way, your soul is at rest."

It was around then that she began to see ghosts.

"They just started appearing in my life," she says calmly, as if she is talking about stray cats. Word spread, and soon she had strangers calling for help with phantoms. "I don't do it to help the home owner; I do it for the ghosts. Some want the ghost to just leave, which can be a problem if they don't want to. Some just want a certain behavior to cease and desist. But then the ghosts tell me what the humans are doing that's bugging *them*."

Bayless moved to Colorado, where she felt her interests raised fewer eyebrows. Ghost busting—she prefers to say she "helps" ghosts—is a part-time job. She earns her living custom-painting home interiors and volunteers as an EMT and firefighter. She helps ghosts in sessions held on-site for $100 to $150 plus travel, or over the phone.

Over the phone? As credible and matter-of-fact as she sounds, there's something a little 1-800-PSYCHIC about telephone ghost busting. Bayless shrugs off such skepticism. "I find it's easier sitting in my own living room. I do it in meditation, anyway."

Bredo Morstoel's ghost came to her in meditation. For all the hoopla surrounding his dead body, Bayless reports, Grandpa Bredo "has a good time with it." I notice she does not add qualifiers like "I think" or "maybe." "He likes the notoriety and the ridiculousness."

I ask if she runs into Grandpa Bredo's ghost all the time, and she sounds surprised. "He only hangs around during the festival and occasionally in the year," she says. "I mean, it's not as if he ever lived here." Grandpa Bredo's ghost, it appears, has other places to go and things to do.

Meanwhile, Grandpa Bredo's body is giving visitors from around the world a place to go and things to do in the town of Nederland, Colorado. In her postmortem account, Barbara Lawlor says Frozen Dead Guy Days 2005 attracted over seven thousand visitors. It made the *Today* show. Perhaps a handful of those visitors learned a thing or two about the science of preserving the human body for possible future reanimation. Certainly a great many had a chuckle or two at the expense of death, lightening by a shade the darkness and terror they'd perhaps previously associated with the subject. They learned that once upon a time, there lived a man named Bredo Morstoel who lived happily but not ever after. His loving but misguided grandson had other plans, which led to a science experiment gone awry.

The end, in any event, was only the beginning, as Morstoel's death allowed a community and its guests to reexamine their attitudes toward the dead. It was the beginning of "a festival about death and the folly of cryogenics," as Amy Bayless puts it. "We're not gonna resurrect somebody's body. If you've been hanging around that long, I don't know that you want to come back."

The Plastic Man

PRESERVING OUR BODIES FOR SCIENCE AND SHOW

Children examine a plastinated cadaver at a "Body Worlds" exhibit.

*S*elf-preservation is a popular American pursuit. We spend so much time while we're alive trying to keep ourselves whole and young and unscarred and unhurt that it seems only natural to continue those efforts in death. For some, this extends no further than embalming, which lasts in most cases a week. For those interested in a more permanent form of preservation, today's marketplace offers modern mummification and cryonic freezing.

Those processes serve little purpose other than fulfilling the wishes of the deceased—not that there's anything wrong with that. But for those bent on preserving their body so that it may forever serve the masses, there is plastination.

Yes, plastination. It's a made-up word, derived from the Greek (*plassein*, to mold), invented, or at least popularized, by a German anatomist in a black fedora. Plastination is what Dr. Gunther von Hagens calls his patented method of infusing the human corpse with a mix of plastics, thereby preserving it.

In recent years, the doctor has taken his work on the road, showing a collection of the whole bodies or body parts of fifty plastinated individuals in twenty-six cities around the world. To date, 17.5 million people have trooped through the exhibit, awed, mesmerized, completely grossed out—or so moved that they signed a form committing their bodies to plastination.

This is why, on a brilliant December afternoon, I have come to stand eyeball to plastinated eyeball with a cadaver.

I am at the California Science Center in Los Angeles to view "Body Worlds," von Hagens's traveling exhibit of plastinated human bodies. In my quest to uncover what attracts people to the idea of preserving their corpses, I thought this exhibit might speak to a motive I had not yet considered: altruism, or something like it. It's not charity, exactly, to donate one's body to science, and "Body Worlds" is not straight science, exactly, but also spectacle. This cadaver I stand before once belonged to a living person who decided that exposing himself this way—stripped of skin, pumped with plastic, and posed like a runner—would somehow improve the likes of me.

Perhaps it would. But the baby, not so much. Mika gives a whimper. My husband, Chris, snatches her and strides toward the exit. I roll my eyes at the cadaver. A glimpse of human jerky won't scar an infant for life. Right?

"Body Worlds" is a singular exhibit: Its subject matter and its content consist entirely of humans. There are glass cases holding organs, some whole, some sliced prosciutto thin, some pink and healthy, some pocked with disease. There's the bright red web of an entire nervous system from brain to toes, extracted somehow in one piece from a body donor. There's the rubbery tube of a gastrointestinal tract from mouth to anus.

The body parts are fascinating. But the exhibit is both famous and notorious for something else: its whole-body specimens.

Von Hagens's controversial innovation is to freeze his whole-body specimens in lifelike poses: running, swimming, playing chess. Rid of fluids, infused with plastics, and stored in gas, the body remains pliable enough to be molded for at least a few months, giving him plenty of time to experiment with just the right pose. Each specimen is meant to exhibit some anatomical detail, each pose to best display its function.

One specimen, for instance, was skinned. It is posed standing with the arm stretched before it, holding the skin like a wet-suit, awing viewers with the reminder that the organ is in fact our largest and heaviest. Less clear is the educational purpose of eyeballs, lips, and hairy genitals left on otherwise stripped-down cadavers—not to mention the cigarette poking out of one specimen's mouth.

A number of the plastinates are captured in athletic pursuits, the better to show our organs and tissues in action. The Runner is supposed to exhibit musculature; his muscles are peeled away from the bone, like ruffled feathers, making him appear more bird than human. The Goalkeeper is suspended in midair, two hands on a soccer ball, his spinal column pulled away from his body to show the intervertebral disks that cushion the vertebrae and allow for strenuous movement. The Basketball Player is frozen in midlunge, left arm extended as if to ward off an opponent, right palm cradling a bright orange ball, mouth open as if to call out a play. The man's defined arm muscles form a thick braid of deltoid and tricep and bicep.

The Chess Player is another of von Hagens's signature specimens, posed in seated position, head and torso bent forward in concentration, right hand curled around a white knight. His central and peripheral nervous systems glow red and exposed, and the top of his skull is sawed off to expose plastinated brain. From the back, you see his exposed spinal cord. Everything, you realize, really is connected.

And then there are the so-called exploded specimens—bodies that have been carefully blown apart to show wrapping and content in vertical order. This one looks like a totem pole, with the face and skull perched like a helmet atop the brain, and so on.

Fascinating though the human body is, at some point you have to remember these are not dolls but real people, real people who *died*—many of them, it appears, in their prime. For me, the most difficult to look at are the pregnant women and the fetuses.

Some of the harshest criticism of von Hagens surrounded a plastinate called "Reclining Woman in the 8th Month of Pregnancy." Its title pretty much says it all, but it was her pose that attracted such angry attention: The woman is lying on her side, one arm propped beneath her, the other stretched above her head, à la *Playboy* centerfold. Her abdomen is sliced open to reveal a completely formed fetus.

It is not so long ago that my own body—and my own, very unplastic baby—looked like this. I feel sick.

At the end of the exhibit is a table set up for museumgoers to fill out questionnaires. Von Hagens and his team are savvy modern scientists who know the value of surveys. Most of the schoolkids who pack the exhibit ignore the forms, but a handful of adult viewers are sitting down, chewing pencils, and carefully considering the questions.

Yes, they write (as I learn later from the exhibitors), 72 percent thought it important that the show consist exclusively of authentic human specimens. Half found the plastinates aesthetically pleasing; 83 percent said they now knew more about their bodies; 45 percent became "more reflective about life and death." Two-thirds felt greater respect for the wonders of the body; and 56 percent felt inspired to lead healthier lifestyles.

The exhibit apparently also packed an emotional punch: Three-quarters intended to devote significant time to dealing with feelings aroused by the exhibit. To measure how deeply the impact stuck, von Hagens's aides conducted a follow-up survey in Vienna six months after the exhibit there closed. A third of respondents said they had indeed improved their diets; a quarter had stepped up the exercise; and 9 percent had given up the smokes—no doubt haunted by the hideous sight of a cancerous lung.

In what the scientists consider the greatest affirmation of their work, one in five said the exhibit made them more willing to donate an organ; and 23 percent said they'd consider bequeathing their bodies for plastination.

That's almost one in four. In London, that ratio was 35 percent. I'm thinking von Hagens is going to find some takers in California, the home state of alternative death.

I'm not mistaken.

BETTY BERDIANSKY
1942–

*B*etty Berdiansky is not plastic—yet.

She is a slender strawberry blonde who loves the jewel tones of fabrics from Southeast Asia and the diversity of West Los Angeles. She gets excited when faced with a brain-twisting logistical problem, which is pretty much every day in her job as computer project coordinator for a city hospital. For her sixtieth birthday, friends threw her a party at a Thai restaurant, complete with Thai dancers and live music and singing. Betty wore a floor-length wrap dress in sky blue and took off her shoes and danced all night.

One weekend, a friend with whom she often visited art museums and theaters invited her to a different type of exhibit. Betty walked into the California Science Center and felt—something. Awe at what she saw as the artistry of the human body exposed. Curiosity at the passionate reactions of the other viewers, especially those of the kids. But what she really felt, she says now, was destiny.

"I knew," she says. "I knew immediately. I was to do this."

To her friend's surprise, Betty sat down and offered up her body for plastination.

Now, Betty Berdiansky was not a woman who spent a lot of time pondering death. Oh, she'd signed up to be an organ donor, but as she drifted into older age, she doubted her organs would do anybody much good. She'd had asthma and surgery on a precancerous colon that had left it truncated, chugging up instead of down. Lately she'd been pondering cremation, her ashes scattered under a tree.

But perhaps she'd make an interesting specimen instead.

That was in August 2004. In September, she got a response to her questionnaire. She and twenty-five other carefully selected potential donors were invited once again to the science center, where they would meet Gunther von Hagens himself.

They gathered in the center, these two dozen, sudden soul mates. A small man appeared, wearing his trademark black fedora. Von Hagens showed them around the exhibit, pointing out his prized specimens and explaining his work.

At the end, he turned to them and invited them all on an all-expenses-paid trip to Heidelberg, Germany, to tour his plastination factory in person.

"We just about fell on the floor," says Betty. "All I could think was, I'm going to see how I'm going to die. My first time in Europe, and it's to see how I'll die."

In Heidelberg, the group toured the charming town, trailed by a BBC film crew. At von Hagens's Institute of Plastination, they received lessons on anatomy over four days. Some involved dissections of corpses. "I'd never seen that before," says Betty. "I would have assumed that would've been upsetting." Strangely, it wasn't.

When the group returned to California, it was as though they had stepped back onto planet Earth after a visit with aliens. "It's like in *Close Encounters of the Third Kind*," says Betty. "We just have this bond." The donors discovered commonalities: Many were educators and artists and health-care professionals; all were independent thinkers with upbeat personalities. They ranged in age from the twenties to the eighties. They were white and Hispanic and Asian; three were African Americans from the same family. They were Jewish and Christian and agnostic. Each felt a powerful conviction that plastination was their fate.

In surveys, potential donors give scattered reasons for offering their bodies to plastination. Some cite an admiration of the artistry; others say they prefer it to the organic decomposition implied by a burial; yet others regard plastination as the ultimate expression of thanks to the wondrously durable body. One potential donor writes: "I'm a Bible-believing Christian and I know that man has a soul, and when he dies, the breath of life within him will return to God. And when Christ returns I will be raised to new life in a new body, as is written in the Word of God."

Betty believes in life after life, too. Raised Presbyterian, she abandoned organized religion in the 1960s, a move she links vaguely to the rampant racism of the times. "I define myself as spiritual," she says. "I have come to that as an adult. I don't have the same holding on to my body that many people would. My identity is not in my body."

In previous lives, she thinks, she has been a Thai prince and a stained-glass craftsman. The body of a sixty-two-year-old Caucasian woman is just her current vessel. This detachment is something she shares with the other donors. "I think we're all comfortable with the idea of death. We're not afraid of it."

That detachment, unfortunately, is not always shared by family. Betty is long divorced but has a grown son and siblings, whom she says are alarmed by her wishes. Her son's concern, she says, has to do not with donating her body, but with putting it on exhibit.

"I have a birthmark on three of my fingers," she says, laughing indulgently. "He says people will see it and know it's me." Then she grows serious. He is her executor and only child, after all. "Do you do what they want because they're the ones living, or do they do what you want? I do consider that he may not carry out my wishes. If he chooses not to, I won't lose a lot of sleep—because I'll be dead."

Privacy is among the leading concerns of potential body donors, judging by the amount of writing von Hagens devotes to the subject in medical papers and articles. Anonymity is crucial, he writes, for "distancing the body from its plastinated counterpart"—in other words, separating the person from the specimen. "It is the only sure way of ending the sense of reverence surrounding the body, i.e., the sense of personal and emotional attachment to the deceased." What he can't do, von Hagens concedes, is take away its human essence. "It goes without saying that every human specimen continues to possess human qualities."

Presuming her son carries out her wishes, Betty has her marching orders ready. After she dies, her body will be embalmed by a local undertaker. The embalmed body will be shipped either to Heidelberg or von Hagens's other facility in Dalian, China. There it will be evaluated under dissection, to determine what parts, if any, can be used for exhibition.

Before all that, there will be a party. "One where everyone laughs and talks and just has a good time," says Betty. She hasn't planned much more than that. But it's a good bet she'll wear her sky blue dress and take off her shoes and dance all night.

⁜

*G*unther von Hagens is a busy man. Every ten days or so he travels between his institutes in Dalian and Heidelberg as well as to the state medical academy in Kyrgyzstan, where he is teaching his techniques to a team of "very eager" scientists. In between, he'll hop over to the United States or United Kingdom or whatever country is holding one of his exhibits.

This day in November 2005, he is wrapping up a short stint in Heidelberg, during which he has visited the possible site of an expanded new facility, arranged permission to export an elephant carcass from a German zoo to his plastination institute in China (which has more space), and begun preparing a Christmas letter to his 6,500 body donors around the world. Tomorrow evening, he will fly to China. Tonight, he calls me at my home to discuss the philosophy behind plastination.

Dr. von Hagens, sixty, is bright and exuberant and funny—not quite my image of the man the German press dubs "Dr. Death." In reedy, heavily accented English, von Hagens recites his oft told bio. He was born in what today is Poland during the waning months of World War II, he says. To flee the oncoming Russian army, his mother tucked her five-day-old baby in a basket and cycled to what became East Germany. There the young von Hagens attended medical school.

One day in 1965, he attempted to sneak across the border. He was caught, thrown in jail, and incarcerated for two years until West Germany bought his freedom for about $20,000.

He attributes his bold pursuit of the unknown to his experience in captivity. Soon after his arrest, he says, a Czech guard in a Bratislava prison purposely left open a window for his escape. Von Hagens hesitated—and missed his chance. After that, "I made a clear decision in favor of uncertainty, because uncertainty represents an opportunity," he writes.

In West Germany, he resumed his medical studies, specializing in anatomy. Von Hagens never shied from new, nontraditional ideas. For instance, watching a butcher slice deli meats in a butcher shop one day, he realized he could obtain deli-thin slices of, say, human kidney using a rotary blade cutter. And when he first saw a specimen embedded in a block of clear, solid solution, he wondered why the polymers couldn't go inside the specimen instead of around it.

His greatest obsession lay in what he calls "the democratization of anatomy"—in other words, finding a way for the general populace to experience the wonders of the human body in a way that only scientists and doctors usually do. Plastination, von Hagens realized, could achieve that. By replacing bodily fluids first with acetone and then with plastics, then curing the body using heat, light, and gas, he could achieve an odorless, rubbery specimen unthreatened by decomposition—one that could be posed and moved and displayed for all the masses to see.

But first, he would need body donors. He began by approaching people who had offered to donate their bodies to the University of Heidelberg, asking if they would willingly consent to plastination instead. But after his exhibits opened, von Hagens was flooded with offers.

So far, 6,500 people have signed on as future specimens. About 45 percent are men and 55 percent women, according to a 2005 survey. A third are currently over sixty, but the age groups are fairly evenly represented otherwise, including 5 percent who are between eighteen and thirty. Four out of ten are Christian.

The vast majority are from Germany, where "Body Worlds" has been exhibited most frequently and where von Hagens is a celebrity. But since launching the body donor program in the United States just one year ago, 150 Americans have signed on. In fact, in a poll sponsored by his institute and conducted by a German market survey firm, one out of ten Americans said they would consider plastination for themselves—before they had even seen the exhibit.

This came as a pleasant surprise to von Hagens. But he understands the appeal. America is "a body-centered society," he says, "with the exercise and the plastic surgery and everything." After a lifetime of concentrating on the betterment and upkeep of our bodies, it's hard to just let go.

"Americans fear that the last time their body will be shown is at the funeral," he says. Thus the attraction to preservation, even if only by embalming. In Germany, he notes, embalming is not even allowed except in unusual cases. Yet our religions do not address what happens to the body after death, focusing instead on the soul.

"The body is the best representative of the soul," he says. "And to be plastinated is a kind of secularized burial."

Plastination also seems to appeal to the individuality and choice that Americans crave. Of all the body donors from nineteen different countries, Americans are the only ones who have demanded that they be posed in a certain way and placed in a certain place. "Even the pharaohs did not get to be posed a certain way," says the doctor.

———

But it turns out I am dead wrong about the motive.

Altruism—at least the kind of altruism that leads a person to donate an organ to a stranger or the whole body to medical research—does not always figure in body donors' minds. Von Hagens says his donors recoil from the idea of giving up their bodies for clumsy residents to hack up. (Betty Berdiansky backs this up: "I've seen medical students at work, and I'll take a pass, thank you very much.")

I had also assumed that donors, while eager to be plastinated, would object to the postmortem fame and exposure of their cadavers. Here, too, I am completely off. No—Americans who choose to place themselves in Dr. von Hagens's hands *crave* that kind of publicity.

Fully half of American donors say they will allow their names and other identifying details to be displayed in the exhibits. Some even request it. " 'I am Harry, and I want to go to my university as Harry, and I want to show effects of smoking,' " says von Hagens in his best American accent.

Offering whole-body plastinates for display in public areas is something von Hagens and his team are currently negotiating. Posing a formerly living human in a university rec hall or a hospital lounge sounds controversial enough. That's nothing compared to what else von Hagens has planned.

In the recent survey of body donors, the doctor posed a series of questions to judge the limits of donors' tolerance for weirdness in the name of science. Would they consent to being posed in a group, such as in a string quartet? Would they allow their bodies to be posed with a plastinated animal, like the horse and rider in the current exhibit? Would they let their bodies be *merged* with an animal—say, as a centaur?

Some of the questions show that von Hagens has every intention of continuing to push society's buttons. Would donors allow their bodies to be crucified? . . . to undergo a postmortem sex change? . . . to be displayed having sexual intercourse?

The responses were mixed. While generally accepting of increased publicity and anatomical experimentation, donors put the kibosh on some of the more outlandish ideas. Centaurs and sex changes were, for the most part, voted down.

And surprise, surprise: Many more men than women agreed to engage in postmortem sex. This was true even among the Viagra set. Talk about the final score. Is it still called necrophilia if you want to do it when *you're* dead?

In the end, I've come to believe donating one's body for plastination is about more than preservation. It's about more than denying nature its course. It's about more than medical altruism.

It's about pride, and curiosity, and gratitude—of, in, and toward the body that endured decades of crappy diets and chemical experiments and broken bones and lousy posture. It's about marveling at the incredible survivor that is the human body and letting others marvel, too. And what better way to thank that body than by giving it a shot at immortality?

If that's you, give Dr. von Hagens a call. He'll be vaiting.

"The Culture Thing"

NEW AMERICANS, OLD FUNERAL RITES

*At the funeral of Yung Her, her Hmong family mills around her
casket as a qeej player guides her spirit to the other side.*

The Hmong Funeral Home of St. Paul, Minnesota, is a squat,
gray-painted building along a stretch of Dale Avenue with auto
repair shops and ethnic beauty suppliers. In the parking lot behind the
building, Asian men lean against cars, pinching cigarettes between their
index fingers and thumbs. Hovering above their heads is a cloud of yellow-
jackets.

The only way in seems to be through the unsmiling men and the
wasps.

I am crashing a Hmong funeral. I had wanted to observe the funeral rites of new Americans: Which traditions did they insist on continuing and which did they bend to suit their adopted homeland? How did the American funeral industry respond to their unfamiliar demands? Why did carrying on their death rituals matter so much to immigrants? Did the younger generation know or care what it all meant?

Of all new Americans, I thought the Hmong, with their elaborate funerals and their intact culture, might hold the most revealing answers.

The Hmong originated in China, and for the last few centuries have inhabited the mountain regions of Laos. Many Hmong emigrated to the United States after the Vietnam War. Their intimate knowledge of the mountainous jungle aided the United States during the war, an alliance that marked them for the hunt by the victorious Vietcong.

Among immigrant groups to reach these shores of late, the Hmong had perhaps the furthest to leap; their animal-sacrificing rituals and guttural language have not contributed to an easy assimilation. Clans have clustered together in California, Minnesota, and Wisconsin, clinging to the old ways in the onslaught of the new, desperately imprinting on their Nickelodeon-bred children the ancient rites of life and death.

Of approximately 186,000 Hmong in the United States, 41,800 live in Minnesota. That, at least, is according to the 2000 census; tribal elders suggest many of their kinsmen distrust and avoid such governmental polling and put the number at closer to 70,000. Brought here by missionary sponsors, they stayed for the state's compassionate social programs and the manufacturing jobs, says Dia Cha, an associate professor of anthropology and ethnic studies at St. Cloud University in Minnesota. The original settlers were soon joined by siblings and cousins until entire clans lived within a few zip codes.

There are three Hmong funeral homes in Minneapolis. Until 1993, there were none. Until then, Hmong funerals were conducted in Western-style funeral homes, causing much bafflement and distress among both the families and the funeral directors.

Jim Bradshaw is one of those business owners. Bradshaw says that over the years he has conducted up to four hundred Hmong services in his seven Twin Cities funeral homes.

When I arrive in Minneapolis, I drive out to Bradshaw's newest facility, a high-tech, übermodern, $5.5 million funeral home in the suburb of Stillwater. His son Jason, also a funeral director, shows me around the spacious one-story building, pointing out the skylights and the webcam and the million-dollar sound system. The walls and furniture are in muted forest colors. There's a toy-filled playroom. The floor-to-ceiling windows in the community room open onto a terrace surrounded by a man-made prairie. Someone even held a wedding here earlier this summer. In fact, there's no mention of funerals in its name: the Bradshaw Celebration of Life Center.

It embodies the modern American funeral ethos. But I don't see a designated room for animal sacrifices.

"When we were first asked to conduct a Hmong funeral, we had no idea of their needs—or they of ours," Jim Bradshaw had said when I spoke to him earlier by phone. The funeral director and the families tussled over embalming, a practice considered standard by most American undertakers but barbaric by the Hmong. Bradshaw had to explain to community leaders the state's requirement that any body not buried within seventy-two hours— the minimum duration of many Hmong funerals—had to be embalmed.

Faced with little choice, they acquiesced but demanded Bradshaw heed their own requirement that nothing artificial touch the corpse. "So we'd lay the bodies out on sheets of natural gum rubber, and any sutures we did, we used natural thread of flax or cotton." The damage to the internal organs in the embalming process also caused great consternation among the Hmong, who believe in reincarnation and thus the sanctity of the body in death. Their outcry was such that Bradshaw and other funeral directors agreed not to remove or otherwise harm organs except in the case of disease.

Integral parts of the Hmong funeral rites, like the animal sacrifices, were trickier to accommodate. Hmong families compromised, letting a local butcher kill the animals instead.

Try as he did to adequately serve Hmong families, Bradshaw understood that their needs in death differed so greatly from Western standards that dedicated facilities would do the job best. Indeed, in recent years as the Hmong have acquired their own resources, Bradshaw has conducted only a handful of their services. He and his son speak highly of the Legacy Funeral Home, a state-of-the-art facility that opened in 2005 and seems likely to attract much business in coming years. But the weekend I am in Minneapolis, its owner tells me it has no services booked.

Determined to witness a Hmong funeral, I try the other two funeral homes. The Hmong speaker at one of the homes repeatedly hangs up on me, terminating my final call with a lot of shouting. I have better luck with the straightforwardly named Hmong Funeral Home. Sue Xiong, the manager's wife, picks up the phone. As I explain my project, she cuts in.

"So you want to see Hmong funeral? You come. I think family is okay."

———

So here I am. I am wearing my usual funeral-crashing outfit of head-to-toe black, but it appears this was unnecessary; guests and family members are dressed in their workaday slacks and shirtsleeves. My Asian appearance may have allowed me to slip in relatively unnoticed except for one dead giveaway: my height.

I am five feet six, which makes me a good head taller than some of their men. The Hmong, being a mountain people, benefited from small stature—in high altitudes, oxygen has an easier time reaching shorter limbs—until they came to this country and had to reach high shelves in grocery stores and conduct business with monsters like me. Conveniently, their American-born, milk-bred offspring have grown to average American height. But the generations born in the old country remain very small. I notice at least two middle-aged men wearing dress shoes affixed with man-heels.

American funerals are, for the most part, open to the public, with death notices posted in newspapers announcing the where and when of the viewing and service. Once, at a Catholic funeral in Lyndhurst, New Jersey, I watched a lady whose parked car had been blocked in by the hearse lope, resigned, into the church for the funeral mass of a complete stranger. "I may as well say a prayer for the poor guy," she said, shrugging.

The Hmong take public funerals to the extreme. There are Hmong newspapers that publish obituaries, and even the English-language *Pioneer Press* began printing Hmong death notices in 2004. Two Hmong radio programs broadcast funeral notices. Neighbors tell neighbors. In the end, anyone in the Hmong community who had a passing relationship with the family may show up. Hundreds do—though few, if any, non-Hmong are among them.

The only way into the Hmong Funeral Home, past the wasps and the smokers, seems to be an open back door. To one side of the door is a large white tent, to the other is an attached garage. As I approach the door, I glance to the left into the tent and see metal vats of rice steaming over open fires. I look to the right into the garage and see a dead pig.

I double-take. There on a table is the biggest, deadest pig I've ever seen. Behind the table is a counter on which rises a mountain of raw meat, hacked into fist-size chunks. It looks to be beef, judging by the color and a not-yet-hacked, cow-size leg propped against the mountain. It is the most raw meat I've ever seen, and that's including the butcher counter at Whole Foods.

Not knowing how to behave in this very foreign environment, I revert to Japanese mannerisms. This has happened before. When I speak to people whose language I don't share—say, Spanish—I inexplicably adopt a Japanese accent. I understand this is a universal reaction; when we visit my family in Japan, my American husband sometimes blurts out something in high school French.

At the Hmong man chopping the meat, I incline my head in a short, quick bow, as I would in Japan to convey respect to a stranger.

He freezes midhack with his cleaver above his head and a cigarette teetering on his lower lip. He stares.

I hurry inside. Beyond the open door is a long, narrow room that looks as if it has been set up for bingo. Three men sit at the head table, shuffling a stack of envelopes. Old men sit in straight-backed chairs at the long tables, watching a taped Cambodian soap.

The room falls silent as I stumble in. I bow. They stare.

Eyes on the floor, I scuttle across the bingo hall and burst into a viewing room, stopping just short of a very colorfully dressed corpse.

The body is laid out on a pallet about knee-high. She—it is an old woman—is wearing a costume of brilliant jewel tones, with a gauzy purple turban on her head and curly-toed purple slippers on her feet. Her face has the waxy, grayish hue of the embalmed. Unlike most funerals I have attended, this one has the feet of the deceased pointing toward the left side of the room. Professor Cha tells me later the Hmong always lay their dead with the feet toward the door.

Two men sit on stools by the body's side, taking turns chanting into a microphone. According to Hmong belief, the soul of the dead must travel through every place of residence it has ever held until it winds up at the house where it was born, under the dirt floor of which is buried its placenta. Only when the spirit dons the golden jacket representing this placenta can it rest. The men chant words encouraging the spirit on its journey.

The room is long and narrow. The building must have been a modest church in an earlier incarnation; this would have been its chapel, with the body laid across the altar. Rickety chairs are lined in tight rows all the way to the back of the room. A smattering of women sit on one side, chatting pleasantly over the din of the large metal fan. There is no air-conditioning, and the August heat is thick. Across the aisle the men sit, legs splayed, arms crossed, nudging one another with their elbows and swapping jokes. A little girl in a blue T-shirt runs by the corpse; her grandmother beckons her to the door with a sliver of peach. An infant is passed around.

To match the generally convivial mood, the decorations, too, reflect a party atmosphere. Gold paper cutouts are strung like streamers across the ceiling. I learn later that these represent money—money the deceased will need to buy herself a home and other material goods once she reaches the other side. In this culture, it seems you *can* take it with you.

On the front wall of the room, surrounded by more fake money, is an eight-by-ten portrait of the deceased when she was younger. She is a smooth-faced woman standing against a brick wall, wearing a long black skirt, a red sash around her middle, and thick silver around her neck. It's the authentic version of the bohemian peasant look so popular this summer; she looks festive and kind of trendy.

Under the portrait is a glossy cherrywood casket. It is empty. The top half of its lid is open when I arrive, at 10:00 a.m. on a Friday, its frilly white insides displayed, a pile of red and white roses on its closed lower half.

After a while, I track down Sue Xiong, the manager's wife, but she makes clear that it is not the funeral home but the family that is in charge. I find this hard to believe, given the scale of the operations. But aside from the embalming, which has occurred in a Western funeral home, the relatives are indeed running every aspect of the service here—the ceremony, the cooking, the money collecting, the pig decapitating. When I pose a few basic questions about Hmong funeral practices, Xiong waves her hands and defers all answers to Dang Neng Yang.

Dang Neng Yang, forty-three, is the stepson of the deceased, whose name is Yung Her. She did not raise him; Dang's father married her after they immigrated to the United States in 1976. She has six children of her own blood.

Still, Dang is, for all Hmong purposes, considered the eldest child. Therefore, he must bear full responsibility for the funeral. He will conduct the intricate ceremonies. He will remain present and (mostly) awake over its seventy-six straight hours. He will scribble checks and peel off twenties to cover what will wind up being a staggering funeral bill.

Dang's eyes already bulge from lack of sleep. Yet he brightens when I introduce myself and ushers me outside so we can talk away from the din. He speaks English fearlessly, blaming a slight slur on a stroke in 1999. His tongue worked better, he says, during the twenty-five years he worked as an interpreter for the city of Minneapolis. In this strange world, I could not have hoped for a better guide.

YUNG HER
1945–2005

*Y*ung Her, like many of her people, had known hardship and drama. The Laos-born mother of three lost her first husband, who fought on behalf of the United States during the Vietnam War. She made her way through the refugee camps of Thailand to the

United States and finally to Minneapolis. There she met a widower with two sons of his own who had lost his wife to disease in the refugee camps. They married. She would bear him three more children. Her eight children would bear her forty grandchildren, who would produce four great-grandchildren by the time she died, four days ago, on August 14, 2005.

Yung Her's spirit is at this moment on a path to heaven, guided along by the chanting. At least, heaven is what Dang calls it—a shorthand he employs for my sake. The Hmong practice a combination of ancestor worship and animism and don't believe in a heaven or hell per se; it is her ancestral home toward which her spirit is heading. Her journey will be long because she was old, Dang says, requiring perhaps seven or eight hours of chanting. For this reason, the older the deceased, the longer the funeral.

Dang isn't sure of his stepmother's age—few Hmong born over there seem to keep track—but after conferring with numerous relatives, he settles on sixty.

"Old," he says.

Maybe, for the Hmong. But life expectancy for women in the United States averaged 80.1 years in 2003, according to the Centers for Disease Control and Prevention. In Japan, women live to be 85.3 on average. I calculate how much longer we would be here for my Japanese great-grandmother, who by dying at 103 outlived Yung Her by 72 percent; my Hibaachan's funeral, by Hmong measures, would last five and a half days.

I ask Dang to tell me about his stepmother, and his response is unsentimental. "She like it here," he says, nodding and smiling as if she is sitting next to him, sipping a cup of tea.

What did she do with her days?

"Friends, cousins, they take her shopping. Harder because she start dialysis fifteen years ago. My sister take care of her, then when sister marry, my two brother live with her and my dad."

Did she enjoy being around the children?

"Yes, she love the grandchildren. She baby-sit them."

Hobbies?

"She was sewing the skirt and shirt for son and daughter. For marriage, wedding. You know."

We walk back into the viewing room so he can better explain the rituals, and there he exhibits a similar detachedness. He skims past his stepmother to adjust the videocamera trained on her body. Conscious of being so close to the body, I bow in the general direction of the chanters, but they glance at me with indifference. Dang strides up to the body and points to the small red and yellow quilts lying across her knees. "She make these, too," he says.

She made the quilts she knew she'd be buried with?

"Yeah," he says. "She make only for die."

The empty coffin is surrounded by large, standing floral displays, one of the few props I recognize from traditional Western-style funerals. There is Hmong writing on the fat red ribbon tying the red roses on her casket.

Koj Nam Kuv Hlub Jshua Koj.

Hmong became a written language only as recently as the 1950s with the help of the French, who in my opinion did a lousy job. Not only does written Hmong employ consonants with unreasonable frequency, it doesn't remotely match the spoken language.

I ask Dang what it means. He contemplates the message. "It say, my dad will love my mom forever."

It is dawning on me that what I mistook for a lack of sentimentality is in fact a cultural comfort with death. The playing of the children, the milling about the casket, the loud talking and belly laughing and beer drinking and pork eating—all speak to an ease I have not previously seen in American funerals. Here more so than anywhere I am reminded that death is part of life.

This is not to say there aren't moments of pitched grief. By afternoon, the mood has changed. Yung Her's body has been moved from the pallet into the casket. The haunting chant has been replaced by the drone of the *qeej*—Dang pronounces it *keng*—a bamboo instrument that seems to function somewhat like a bagpipe. The players dip and swoop, knees bent, circling the room in slow motion, filling it with an unbroken bleat. One *qeej* represents Yung Her's spirit; the others urge her on her proper path. A drum is strung from a bamboo tepee, tapping out a rhythm that quickens each time a new visitor enters the room. Then the *qeej* players swoop into action, covering the guest's tracks to remind her spirit to keep going despite the company.

Yung Her's blood relatives are arriving, and the emotions are swelling along with the music. Her surviving siblings rush past the *qeej* players and throw their arms around the casket, their wails drowning out the drone of the pipes. They stand around the casket for a long time, crying openly.

I spy Dang in the front of the room, embracing an aunt. His back heaves. When he pulls away, his eyes are streaming.

❖

*W*hen the chanters wrap up, Dang hands each of them an egg in a rice bowl. The gift is to ensure their souls don't follow hers to heaven, he says later.

It turns out the eggs are but a tiny fraction of what he will pay for these four days.

Each of the dozens of helpers must be paid anywhere from $50 to $100 for services ranging from boiling the beef to playing the *qeej* to hauling the cases of beer. Every time a guest walks through the doors and brandishes an envelope, a portion of their monetary gift must be returned. Some will give $10; his sister is driving all the way from Detroit so she can hand over the $2,000 she saved on plane tickets. If a guest steps into the viewing room and the family is not there to greet him, he is paid money to make up for the insult.

All told, Yung Her's funeral will cost $30,000, 90 percent of which Dang will shoulder.

Money plays a central role in the ceremony. Dang is currently hanging what looks like a giant paper macramé from the ceiling. It, too, is "money" from Yung Her's descendants. It will be burned at her graveside on Monday; any paper that fails to catch fire will represent good fortune she bequeathes the living.

The loudspeaker crackles. Every guest handing over an envelope is announced over the microphone by the men sitting at the head table in the bingo room. At the first scratch of the mike, Yung Her's male relatives rush in and surround the donor. They begin to chant, bowing from the waist and swinging their arms forward in a scooping motion. Then the men touch their knuckles to the ground and drop to their knees. The guest pulls them up gently.

When I ask what they are chanting, Dang answers, "It's the culture thing." This is his invariable response to nearly all of my questions. Giving money is the culture thing; sacrificing Porky is the culture thing; the oldest son's obligation is the culture thing. He is not trying to dodge the question. Culture, to him, means tradition and needs no further explanation.

Dang leads me toward the back of the building into the kitchen. There, stout women in aprons lean against metal counters. For the duration of the funeral, they will cook like crazy twice a day to feed the hundreds of guests. The women, having just finished a sprint, are sucking on beef bone piled high on Styrofoam plates. Giant fans blow away the flies. There's a tangy smell wafting off the steaming beef broth, an herb I can't identify.

The serving of the food is occasion for more ceremony. After a hard day of memorializing, the elders have gathered in the building's basement. They sit along one side of a long table. The male relatives of Yung Her stand facing them in the middle of the room. They josh and laugh with the elders, who sit beaming with their hands on their splayed knees. The younger men clomp up and down the stairs carrying aluminum trays sagging with pounds and pounds of cooked meat. The trays are laid out in front of the elders, creating the curious tableau of a row of small people facing a literal mountain of food. They could invite the entire clientele of the Wendy's up the street, and still there's no way they'd finish this feast.

One relative of Yung Her begins pouring from a twelve-ounce bottle of Budweiser into Dixie cups set before the elders. He spears a hunk of beef onto a plastic fork for each of them. There is much cajoling and playful refusing until finally the beer and beef are downed. Then the elder at the end of the table stands up. He is Dang's cousin, Vang Dae Chang, one of the men who chanted for hours by Yung Her's body. Now it is her descendants' turn to chant for him. They sweep their arms forward, bodies bent, as if shooing a toddler. Then they place their knuckles on the floor and kneel.

The elders look extremely pleased. Whatever they are in their American incarnation—security guards, factory workers, truck drivers—here among their own they are given the respect accorded by their familial rank and seniority. The funeral is occasion not just to remember the departed, but to remember their own rightful place in society.

It's the culture thing.

———

The culture thing may be scattering amid the whirl of assimilation, but here in the Hmong Funeral Home at the funeral of Yung Her, it is alive and well and pressed upon the young. There are so many young people that in certain parts of the building it looks like a house party. In fact, 56 percent of Hmong in the United States are under eighteen, according to the census; their median age is sixteen.

Girls in platform flip-flops and low-slung jeans hang around the kitchen, filling plastic cups with Pepsi. "I was like, Don't even call me," one says to another.

The boys seem to have more duties. They haul ten-pound bags of rice on their shoulders and hand bottles of Bud to elders. When a new guest arrives, they hustle into the aisle of the viewing room, where they crouch on their knees, holding two joss sticks in one hand. After the guest helps them up, they disperse outside to smoke and await their next task.

I'm impressed by the display because I am unused to seeing young men performing family functions without goading or, for that matter, at all. I've seen young men at Western-style funerals, of course, lurking around the back, pulling at their starched collars and looking deeply uncomfortable. Here, though, are young men and boys with shaved heads, chains and cell phones clipped to their jeans, kneeling unbidden before their elders.

One young man looks both tougher and more mature than the rest; he is dressed today in a checked button-down shirt untucked from his chinos, an outfit somewhat incongruous with his shaved head and shuttered face. He is first to kneel in the viewing room at the feet of a just arrived guest; he sits at his elders' elbows, holding the envelopes; he jumps up to pay a deliveryman.

"Uncle," he says softly as we pass, "the guy needs the receipt."

Chris Yang, twenty-eight, is Dang's nephew and Yung Her's oldest grandchild. I ask him to talk with me in the bingo room, and to my surprise he readily agrees, though it quickly becomes clear that he is a man with boundaries.

Chris is married to another Hmong, with whom he has a two-year-old boy. I ask where they are.

"Not here," he says.

He works in production for Dow Chemical. When I ask him to be more specific, he stares at me. "It's manufacturing."

Despite his current respectable station, I'm not entirely surprised to spot the tattoos smudging his knuckles, and like the rude stranger I am, I ask if he was in a gang.

"Yeah," he says. "I don't want to talk about it."

Eventually he opens up. His branch of the Yangs arrived in the United States in 1975; he was born a year later. Longing to work the land, Chris's father moved his sons from Minneapolis to California, where they owned a small farm that grew cherry tomatoes and sugar peas and bitter melon. "Fifteen acres—that's a lot of freedom," says Chris. "It kept me out of trouble." Freedom didn't last. Trouble found Chris when the family moved to Los Angeles. Chris joined a gang. He and his father fought.

They returned to Minneapolis when his father grew ill. There Chris encountered a culture he hardly knew or comprehended. "Half the time, I don't know what they're doing," he says, gesturing toward the elders. "Sometimes I can't even speak Hmong right, if it's, like, too high of a word."

It's easy to remain lazy students of family history when we're tethered to it by our parents. They are our ancestral cheat sheets, our cultural decoders; what we don't know or can't recall, they'll sigh and tell us (again). I remember when my mother, after a particularly nasty stretch of chemo during her first bout with cancer, showed me (again) how to cook mackerel. Pressing the miso through a sieve to rid it of lumps, adding a sliver of ginger to quell the fishy smell—all these were oft repeated details I never bothered to absorb, knowing they were but a phone call away. I realized just then that the cancer might cut that cord. For the first time during a lifetime of cooking lessons from my mother, I took notes.

Chris's father had cancer, too, and he was dying. You will have to play my part, he told his oldest son, in the family, at funerals, at community gatherings. And his father had one last request.

"I didn't plan to marry yet," Chris says. He was in his mid-twenties and had been dating a Hmong girl for two years. You are the oldest, his father said. If you marry, I will know you can take care of the others. Chris understood. "If I was going to make the ends meet—we always fought, and I just wanted him to forgive me. I didn't want him to go without knowing that I loved him. That I honored his wish." The wedding was an elaborate Hmong affair attended by hundreds, at the end of which his father told him, I am happy. I don't mind going now.

At his wedding, Chris had followed blindly as elders guided him through the complicated rituals his father was too weak to show him. They would do the same at his father's funeral.

"After he died," says Chris, "I realized."

Family swarmed in from every corner of the city and country. A dozen elders conferred on each decision. Every Hmong that Chris's father had helped—no small number, for he had last worked for the city as a social worker—came to pay respects.

"I didn't know," Chris says, his voice low. "I didn't know it would be like this. They helped us, and they kept helping us."

America—with its twenty-four-hour TV, its moneymaking opportunities, and its aggressively welcoming churches—would be too beguiling for many young Hmong to resist embracing. About a third of U.S. Hmong have converted to Christianity, according to Professor Cha. The old way involves a too high language and baffling rituals and the slaughter of farm animals. What could these demanding and superstitious ancestors offer that America could not?

One could even argue that the alienation begins at birth, when American hospitals discard the placenta, leaving Hmong mothers nothing to bury under the eaves of the house. Without the placenta to guide them home, how can the spirits of their American-born children find their ancestral roots?

For Chris, these legends and superstitions and passed-down beliefs meant little until the death of his father.

"I believe now," he says. "I believe in my ancestors. After my dad passed away, I started to realize all this stuff. I was born here. My son was born here. But it's the most important thing now that he knows, too—that he believes in our ancestors and our family. If we didn't, we would have given up a long time ago."

At the head bingo table, a relative picks up the microphone. Another guest has handed over an envelope, and he must be thanked.

"Excuse me," says Chris, and he returns to his people.

Denial Is a River

MODERN MUMMIFICATION IN SALT LAKE CITY

*This one is a prop, but real mummies are made
in this pyramid—for a mere $63,000.*

"Meet me at the pyramid."

I am in Salt Lake City to learn about modern mummies. To be specific, I am here to meet Corky Ra, leader of the only organization I could find in the United States that offers to mummify people. Ra has invited me to his pyramid, the mention of which leads me to envision something grand, something very Egypt or Vegas.

It is not quite that.

For one thing, it's hard by Highway 15 in an industrial patch of the city.

For another, it's not that big. The pyramid is small enough to fit on a one-acre, fence-encircled lot that also houses the 1857 cottage where Ra lives, as well as a building that serves as the office slash mummification lab, as well as another shed where Ra bottles wine, and out back a Mini Me pyramid covered in wood shingles. The whole setup makes me think of a really quirky Lego village.

Ra, sixty, is dressed in Levi's, white Fila sneakers, and a green wind-breaker, his gray hair in a ponytail. Being that he's a white man from Salt Lake City, Ra, of course, is not the surname to which he was born. In 1980, the former Claude Nowell legally adopted the name Summum Bonum Amon Ra, which means "sum total of all creation." He says I should call him Corky.

The main pyramid is covered with copper siding. It has no windows, and it appears to have no doors. Ra tells me it's built in accordance with the mathematical ratio of phi (not pi); at its apex the main pyramid is twenty-six feet high—about two stories—and its base sits forty feet wide.

He presses a switch, and a hidden panel opens up slowly from the side of the pyramid, like the door on a DeLorean. We have to duck to get inside.

"Good," says Ra, "you brought a jacket." It's chilly in the pyramid, the better for meditation, mummification, and winemaking. (The three activities, it turns out, are not at all unrelated.)

The first thing I notice is the dog. A Doberman named Butch guards the entrance, in the way a gargoyle guards a building: symbolically, because he's dead. Butch is encased in a Doberman-shaped mummiform complete with pointy ears and plastered with gold leaf. A bronze plaque on the platform upon which he sits says he lived from 1975 to 1987. At his feet lies a collection of smaller critters in various states of mummification: a parrot, a cockatiel, a rat, and a sparrow named Spike. Strewn about like that, they look as if they could be Butch's chew toys.

At the head of the pyramid (if there is such a thing) is an altar—a long table covered with red lacy cloth and festooned with candles, crystals, and bouquets of peacock feathers. On either side of the altar are human mummiforms: one closed and painted in the classic Egyptian style, the other open with a bound-up, person-type shape inside.

"That's a media prop," Ra says cheerfully. He knocks it with his fist. "See? Hollow." A dummy mummy.

The three cat mummiforms at the dummy mummy's feet aren't fakes. They contain kitties that once belonged to Ra: Oscar, Smoky, and Vincent. If we broke open their seventy-five-pound bronze coffins, chipped through the resin, and unwrapped the layers of gauze, inside we'd find three perfectly preserved felines, their tails and paws and whiskers appearing exactly as they did the day they died.

Ra and his team have mummified 1,200 animals, most of them pets. In fact, behind the sofa where we sit are three blue-and-red plastic coolers, the kind you bring to a tailgating party. Each contains a cat.

A yellow Post-it note on one cooler reads, "Filled to lip with alcohol and Shura." I'm guessing the alcohol is not beer and that Shura is not a type of cheese.

<hr />

But what I want to know about is the humans. There's a steel vat in the pyramid that looks about the size of a metal casket, but Ra says it's for fermenting the sacramental wine. I am a little disappointed to learn that he has yet to mummify a paying customer. Even with the $63,000 price tag (not including $50,000 to $500,000 for a mummiform), Ra claims 147 people have prepaid for mummification. So it's not for lack of interest; it's just that no one has yet died. That tells me many mummy wannabes are young. Ra says he has fielded inquiries from celebrities. No, he says, he cannot tell me who.

While researching and trademarking his mummification method, Ra and his team did successfully mummify thirty humans, most of them "indigents" abandoned at the city morgue. After mummification, their bodies were cremated in accordance with the law. But five mummies—private citizens who had donated their bodies for research—are interred in mausoleums around Salt Lake City. No, he says, he cannot show me where.

That any of this should exist in the state capital of Utah is in and of itself a wonder, as Utah is nearly 70 percent Mormon and the followers of the Latter-Day Saints most assuredly don't approve. Ra himself was born LDS (the preferred shorthand here); the son of a successful Salt Lake City developer, he lived most of his life off a trust that freed him to pursue philosophical interests instead of paying jobs.

In 1975, following a failed marriage, Ra says he experienced an "awakening, a personal inner transformation." It involved an alien visitation, but that's another story. He realized his "essence" was eternal—that when he died, it would persist and return. He left the Mormon Church to create Summum, defined by Ra as a philosophy and by the IRS as a religion. (He paid for his betrayal; the hefty inheritance due him was given instead to his daughter, he says, by his scandalized LDS sisters.)

Ra realized he had a lot in common with ancient Egyptians, he continues. "The Egyptians spent their lives preparing for death," he says. The point of life was to live in such a way that death would transfer the soul to some higher state of being. The Egyptians mummifed their dead so their souls would have a form to return to in the next life. Modern Americans, Ra says, could do the same.

"What mummification does is it allows the body to be preserved without it changing or decomposing while the individual is going through the transference," Ra explains—transference being the process by which the person's soul, the essence, transfers to its next state of being after death. Transference is the crux of Summum's and many Eastern funeral rites (much of Ra's beliefs are based on and resemble Buddhism); it's what Ra and his followers spend their lives preparing for.

What's unique about Summum's way of death is not so much the mummification, though certainly that's pretty unusual. It's that the funeral is for the dead, not for the living.

The Summum way of death is as follows. A person prepares for death by writing a spiritual will, a sort of wish list for the next life: deep wisdom, world peace, great hair. The will can contain scripture or poetry or plot points from *Star Trek*.

After death, the body is laid out in the pyramid. A ceremony is performed by Summum members. The altar candles are lighted. Ra tells the dead they are dead ("they don't always know"). Sacramental wine is drunk, with a little poured into the vat for good measure. The spiritual will is read.

The ceremony is so focused on the dead that family members, while not discouraged from participation, are thrown out if they begin to sob. "Grief interferes with the journey," says Ra.

Over the next seventy-seven days during which the transference will occur, the dead person is visited daily and read to from the spiritual will ("they don't always remember").

They are accompanied, in other words, to the other side.

⎯⎯⎯

Meanwhile, the mummification proceeds. The man in charge of mummy making is Ron Temu, whose adopted last name means "worker on the path of creation." When I meet Temu, he is wearing a dark pin-striped suit and glasses; he is tall and tanned and bald. If I saw him crossing Temple Square, I would have guessed he was a banker or some LDS pooh-bah. In fact, he is a funeral director, the manager of nine Salt Lake City funeral homes and a cemetery.

Like Ra, Temu, fifty-seven, was born LDS. For a while, he followed a path the church would approve of, studying electrical engineering and running track at the University of Utah, then entering a Mormon seminary. Then he moved to California. There he tried on different lifestyles. He grew his hair and dabbled in art school; he worked on a freight dock and became a Teamster; he married, traveled, owned fancy cars and a house in Irvine.

All along, he devoured books on philosophy. When he heard Ra speak, his life changed. Temu went back to school to become a mortician, specifically so he could perform mummifications. He has mummified over one hundred animals and participated in the process on humans.

I ask Temu what happens when a body is mummified.

After death, he says, the body would be shipped to Summum, where it is scrubbed clean in preparation. In a modern embalming, the deceased's organs are aspirated—sucked clean of blood and other fluids—from an incision near the navel using a sharply pointed device called a trocar. The mummifying mortician would instead cut "from the sternum to the pubic symphysis" and actually remove the organs. They are cleansed of gunk, placed in a plastic container, and returned to the body. The incision is left open.

The body is then placed in a steel tank filled with a special preservation fluid, the exact makeup of which is a trade secret but which contains no formaldehyde and includes "chemicals used in genetic engineering." (Formaldehyde, the compound used in regular embalmings, "stabilizes the organs for a period but shreds them, too," says Temu. It can also dehydrate the skin so dramatically as to give the appearance of third-degree burns, which is one reason undertakers apply a thick layer of cosmetics to corpses.)

The body thus soaks undisturbed for over a year. "I check it periodically," Temu says. Once thoroughly saturated, the body is ready for wrapping.

Here's the remarkable thing: When pulled out of the vat, that body will look exactly as it did the day the person died. Temu shows me snapshots of Sasha, his Samoyed, who was mummified. After two years soaking in the solution, Sasha looks like Sasha, albeit sopping wet and asleep.

Pulled from the tank, the body is dried off and doused in lotion (preservation fluid, even that without formaldehyde, can be drying). The abdominal incision is sewn back up, organs tucked safely inside. The body is strung up using pulleys so it can be wrapped in layers of clean gauze. It's at this point that it would most resemble horror movie mummies.

A membrane of polyurethane, then fiberglass, and then resin is poured over the body to seal the gauze. Amber resin is poured into the mummiform to freeze the body in place. The lid is closed and welded shut.

Finally, the mummiform is placed in a designated spot—a mausoleum, a "mummy sanctuary," or theoretically even a living room—where its human contents will remain unchanged, says Temu, for the next one hundred, one thousand, or one hundred thousand years.

I wondered if Temu, as a funeral director, displays a mummiform in his casket showrooms or slips in a Summum pamphlet among the bereavement books. He doesn't.

"My job is to sit across from thousands of families and listen to them," he says. "I never talk to them about mummification. I never even bring it up."

That doesn't mean he thinks families wouldn't benefit from adopting

Summum's death rituals. I ask him how he reconciles his work as a funeral director, conducting services for the living, with his participation in Summum, where funerals are for and about the dead.

It's clear he's thought a lot about this.

"Modern funerals, I'm sorry to say, have robbed us," says Temu, leaning forward, pin-striped elbows on his pin-striped knees. "My feeling is that if funerals in our culture were more for the dead—if the individuals left behind were allowed to participate in the process—then I think we would have a better chance to grieve."

At his funeral homes, he lets family members dress the dead—knot the tie on Dad, drape the necklace on Mom. "That little bit helps them," he says. "We don't just drop the dead off. That assists the living greatly not only in letting go, but in feeling comfortable with mortality."

SU MENU
1950–

Su Menu is a fifty-four-year-old piano teacher who lives in Salt Lake City. She is pretty and trim and youthful, with bright blue eyes and a blondish bob. We meet at her town house, where she introduces me to her animals, who have gathered to check me out: four cats, a cockatiel, and a parakeet named Montu, who flaps up to demand, "What are *you* doing?"

"And that," says Menu as we sit down in her front room, "is Maggie."

Sitting quietly amid a jungle of large houseplants is a poodle. Rather, inside a waist-high bronze statue of a poodle is a poodle—a standard poodle, white, dead.

I peek around, but in every other way this is an unremarkable room: flower-patterned sofa on thick green carpeting, an upright piano, framed landscape paintings on the walls. Small children learn piano in this room. Few even ask about the poodle.

Someday, Menu will join Maggie in the mummy world. The rest of her pets will also be mummified and reside alongside her in a crypt underneath Ra's pyramid.

What I want to know is why. After the first five minutes of conversation with Corky Ra, during which he told me about an African tribe of breast-feeding men and how he once experimented with female hormones that "gave me big titties and everything," I had no trouble believing he was a man fixated on a very alternative way of death. But Menu seems, well, normal.

Susan's father moved her family around a lot growing up, following his job for the meat-packing company Armour and Co. The one constant in their life was church. Her devoutly Christian parents were "religious about making us attend," she says. She didn't protest back then; she was focused on the piano. While attending Drake University in Des Moines for a degree in music, she says, "I was sitting in church one day and thought: What am I doing here?"

During her senior year, she saw a flyer for a session on transcendental meditation. "I thought, Well, maybe at the least it would help me with my performance anxiety." Her parents regarded her new interest with suspicion. "My mother asked, 'But do you still believe in Jesus?'"

She met her ex-husband at a meditation workshop and followed him to Salt Lake. Then, in 1976, she attended a lecture at the University of Utah given by Corky Ra.

"I was very skeptical, of course," she says. "Mummification and those things—I was like, Uh-huh. But over time it all began to make a lot of sense to me."

She liked the concept of a spiritual will and the prospect of being guided into the next life.

"It's like having people sit shivah—is that how you pronounce it?—for you," she says. "You don't find many forms of funerals like that unless you're in the upper echelons of the church, like the pope or something. I really like the idea of being helped along."

She joined Summum, and when her beloved poodle, Maggie, died, she had her mummified—not as an over-the-top memorialization, but rather to ensure the dog's company in the next life.

All the arrangements for her own afterlife are settled. The $63,000 will be paid via an insurance policy, the spot in the as-yet-unbuilt crypt under the pyramid reserved for her and her pets. She wants a steel capsule for a mummiform, perhaps with the seven tenets of Summum inscribed on it.

As for a traditional funeral service in which loved ones come to mourn her, Menu scoffs. "You know how people at funerals get up and talk and talk forever—like the person was a saint or something? Oh, what*ever*." She doesn't want that. "Funerals might make people feel better, but when you think about it, when you die, who are you with? You die alone. You are alone."

The thought doesn't frighten her. In fact, she'd prefer it. "I think sometimes, people, their wailing and mourning, can be a distraction. I think you are totally aware. That you're more aware in death than you are in life, because you don't have this body and all these things on you."

She shrugs, suddenly a little self-conscious. "It just depends on if you believe this stuff."

I ask what she has put in her spiritual will. Her current life, after all, does not seem so bad: She has what I presume is somewhat enjoyable work; she has a live-in boyfriend, a policeman; she has two grown sons. But she wants changes.

"I don't think I want children again," she says slowly. "I enjoyed my boys, of course. But I was a single mom. We divorced when the oldest was five and the youngest three. I had to provide for them, and I got very tired of it. I felt like also maybe it wasn't fair to them, especially the older one. It affected him so much."

I ask if that doesn't have more to do with the divorce. If one can dictate such things in the next life, why not just ask for a better marriage?

Meanwhile, one of her cats, Neelix (named after a *Star Trek* character), jumps on my lap and begins to box my belly.

Menu laughs fondly. "I'd rather have cats," she says. "They're much easier to get along with."

I ask what her two sons think about their mommy as a mummy. As next of kin, they would be within their legal rights to rule out the mummification and order a proper coffin-and-grave burial (Utah allows the appointment of a "designated agent" to carry out the deceased's wishes, but relatives would probably win in court). To protect against ugly conflicts, Summum asks family members to sign notarized releases promising not to interfere.

Menu's two boys have signed, as have her parents and two of her three siblings. But one sister, a born-again Christian, has refused.

The thought of posthumous interference upsets Menu. "It's *my* rite," she says. "It's frustrating that someone can take that away. I mean, it's the passage of my soul we're talking about."

Menu knows that all she can do is hope that by the time of her death, attitudes will have changed. "Death used to be taboo," she says. "And now it's not so much. So maybe by then, mummification won't be so scary."

❖

i, everyone! My name is Mummy Bear and this is my homepage. Do you like Mummification? I sure do. I think it's cool. Especially the new Modern Mummification that my friends at Summum do. I'm going to go on a journey to unlock the secrets of Modern Mummification. Would you like to be my friend and come with me? Yes? Then let's go exploring by clicking the links on the left. We'll go on an adventure, and together we'll learn new things and have some fun!!"

If you Google "mummification," a link leads to Summum's extensive and dazzlingly designed Web site. The top of its main page lists a link called Summum Kids. Intrigued, I click and come upon Mummy Bear.

Mummy Bear is a teddy bear mascot intended to educate schoolkids about mummification. Apparently, there is quite a demand for the actual stuffed Mummy Bear, which is shipped to grade-school classes across the country. It's got an incision in its belly through which its organs can be removed.

I am totally creeped out by Mummy Bear. Maybe it's the flurry of dismembered Mummy Bear heads that follow my cursor around the Web page. Or maybe it's the songs and poems. Here's one called "Me and My Mummy Bear":

Me and my mummy bear
Have no worries, have no cares
'Cause me and my mummy bear
Just play and play all day.

He's wrapped up so pretty
And I can unwind
His neat little ribbons,
And then I find—

His tummy comes open
And what do I see?
Special little organs that
Belong to you and me.

I wash them all off
So he can be clean
Then put them back in
And do up his seam.

Then just like a mummy,
I wrap him up tight
Then I cuddle him close
And hold him all night.

Even creepier is the Mummy Bear Prayer.

Now I lay me down to rest
I leave this life, I've done my best.
Please clean my body, head to toe
Wrap me up and make me whole.
Then as my spirit body roams
I'll have a place to call my home.
My body that I lived in here
Will still be there, no need to fear.
Forever now I'll feel so blessed
To have a place my soul can rest.
Amen.

I don't think I want my daughter playing with Mummy Bear. But I do admire Summum's refusal to shy away from death or to shield our children from it. Like many of the world's biggest religions, it asks of its followers blind faith in the rewards of life after death. Yet few of the world's faithful actually embrace death.

"In this culture, people have a lot of beliefs but no one really believes," says Ron Temu. "When you die, you go to a better place, right? People say that all the time. But no one wants to go. They'll do anything, spend anything, to avoid going. Me, I say come on, Dr. Kevorkian. When I can't dance anymore, I want to move on to the next dance."

Seen that way, the fixation on death I encountered among Ra and his compatriots seems less morbid than mystical. The focus on the dead in the funeral rites is also perfectly in line with those beliefs.

I get that. What I still don't get is why the mummification. Sure, I understand the concept of wanting to keep a body intact during the rather long transference period of seventy-seven days; with regular embalming, a body can begin to decompose in one-tenth that time.

Still, why preserve a body so thoroughly that it will keep for hundreds of thousands of days? If the hope is that your soul transfers into a Dalai Lama or an Angelina Jolie, why would you need your wrinkled old shell still hanging around?

The answer seems to lie in science. Ra's followers aren't kooky enough to believe a mummified body will come creaking back to life, but they do trust that science will advance to the point that people can be cloned. Mummification, they claim, perfectly preserves the cells needed for such a procedure. But there are easier ways to preserve cells. I've come across dozens of outfits that offer to preserve the DNA of the deceased.

Despite Ra and his friend's embrace of death, it seems to me that mummification is the ultimate in death denial. Like the ancient Egyptians, these modern Americans wish to hold tight to the lives they led, to preserve the exquisite evidence. We are all of us a death-denying people, what with our open caskets and memorial movies and diamond cremains. The difference is that we, the living, deny the death of our beloveds. By mummifying themselves, Ra and his followers deny their own deaths.

Summum's Web site plays to the vanity of prospective mummies, calling it the only process "that allows you to be memorialized for eternity," the "only form of Permanent Preservation," a rite that lets you "leave this life in as beautiful a manner as possible."

In ancient Egypt, mummification was an option reserved for the elite, and at $63,000 and up, it remains so today. Good thing; just imagine if we mummified everyone who died. Demand would be so high that they'd have to build mummification factories, with teams of surgeons removing organs and rows of vats pickling bodies. Mummies aren't usually buried, so we'd have crypts resembling airplane hangars, jam-packed floor to ceiling with nonbiodegradable storage. There'd be a multibillion-dollar industry in mummiforms, offered in rainbow colors and with compartments for letters and cigarettes. Pyramids would take the place of local funeral homes, causing big headaches for the town zoning board.

As much as we fear and deny death, we are also a society of disposers. We're used to throwing away the things that are no longer useful to us. Sure, we treasure a lock of hair, we honor an urn of ashes.

But I'm thinking it would take a seismic change of habits for Americans to preserve our bodies for all eternity.

Modern Undertaking 101

NOT YOUR FATHER'S MORTUARY SCHOOL

Millennial morticians-to-be at Mercer County Community College.

bout four blocks west of my office in Rockefeller Center is one of the country's oldest mortuary schools. The American Academy McAllister Institute is on the sixth floor of an office building with a Ford dealership on the street level. There's no sign or anything else that betrays its existence. The men and women sucking on cigarettes outside are of an age and a look that say overworked techie more than junior undertaker. I am sure I am in the right place only when I step into the lobby and see a display of granite tombstone samples. At the end of the hall is an open metal casket with frilly blue lining.

I have tried to get inside for a year. Meg Dunn, the school's president, never returned my calls. I finally track her down at the funeral directors convention in Nashville, where the school has a booth manned by Dunn herself. I park my baby carriage in front of the booth and wait, long enough for the merchant across the way to steam-clean my rings for free (his steam machine also "makes those dingy metal caskets real pretty," he tells me).

Finally, an alumna leaves and Dunn looks up. I must sound a lot less creepy describing this book in person—how sinister can the mommy of such an adorable baby be?—because she agrees to let me attend classes and interview students come next semester.

Dunn—Ms. Dunn to her students—is wary with reason. As Paris Hilton might say, mortuary schools are hot. In recent years, as death took a leading role in American pop culture, interest in entering its business has spiked. What Dunn sees more often than she would like, though, is morbid curiosity disguised as career-seeking, exam-taking interest. As guardian of her school, she wants to keep flame fanners like me at bay. She already has dealt with her share of obsessive HBO fans and cape-wearing death groupies and one disturbed young man with a tattoo of the incision point of a trocar.

Some of them, including the trocar dude, even bother to matriculate, which seems to me an awful lot of paperwork just to see a dead body.

To avoid being lumped with the Draculannabes, I explain to Dunn why I am interested in mortuary schools. Funeral service is changing in America, and I want to see the roots of the revolution. For one thing, over half of current mortuary school students are women, and a third are racial or ethnic minorities. This means the face of funeral directing—once white, male, and heir to a family-owned home—will in the near future look dramatically different. Another shift, in no small part affected and even driven by this demographic change, is their focus. Today's student is tomorrow's funeral *planner*—event organizer, cultural arbiter, certified grief counselor.

I want to meet this millennial mortician-to-be.

I tell Dunn I chose AAMI for its sterling reputation and because it seems representative of national trends. This is true. It is also the only one of fifty-four accredited mortuary schools in the United States within reasonable commuting distance from my house.

AAMI is in fact the merged product of the McAllister School of Embalming, opened in 1926 by the anatomist John McAllister, and the American Academy of Embalming and Mortuary Research, a rival institution born in 1933. Since its 1964 reincarnation, AAMI has trained ten thousand morticians. Dunn boasts of a 97 percent pass rate since 1995 at the licensing exams.

A majority—52 percent—of AAMI's student body is female, a statistic that, according to the American Board of Funeral Service Education, exactly matched the 2003 national average. A 2005 survey by *American Funeral Director* magazine puts the number even higher, at 64 percent. Back in 1971, 95 percent of mortuary students nationwide were men.

⟶

One wintry evening, I arrange to meet a majority member of the AAMI class of 2005 at a packed Manhattan Starbucks. A pretty, high-heeled brunette is standing by the cookies, but she looks more *The Apprentice* than *Family Plots*. Tracy Lee, twenty-two, quickly establishes herself as a nice Jewish girl from Long Island. So what was said nice girl doing with embalming juice on her shoes?

Nothing about Tracy's past—or her present, for that matter—shouts mortician. Her father works in marketing, her mom in sales. She shares a two-bedroom apartment in a doorman building in midtown with her sister, a financial restructuring specialist. She's got a bachelor's in criminal justice from John Jay College. She's not even wearing black.

Just before graduation a year ago, Tracy found herself studying for the LSATs with no particular drive to become a lawyer. Then her mother remembered that Tracy's grandfather's neighbor growing up was a funeral director. From that random recall was a career epiphany born.

But Tracy had never so much as stepped inside a funeral home before. So she arranged a tour of a Jewish funeral home in her hometown, "to see if I could handle, you know, being around all that," she says. "Turns out I could not."

The tour began smoothly enough. She met the staff, who began to guide her around the facility. On their way from the lobby to the prep room, Tracy spotted a gurney. She swooned. Her falling body lurched toward the gurney, which had already been blessed by a rabbi. A woman's touch would negate the blessing. The funeral directors shrieked.

Then she passed out.

"When I woke up, they were all gathered around me," she says. "They said, 'Don't worry, you can become an accountant, a teacher—you'll find something.' They were very nice. That's when I decided: I'm doing this."

Tracy sighs. "I didn't even see a *body*."

This episode illustrates the concerns industry elders express at the influx into the business of young women and men with no family connection. They worry that the flurry of pop cultural interest in their profession is attracting the merely curious and the dangerously unprepared. I don't think this is just speculation; I met a young woman who took an internship at a New Jersey funeral home because she liked the A&E funeral home reality show *Family Plots*.

But what they lack in preparedness, they surely make up for in enthusiasm—that is, if it lasts. The mortuary sciences program at Mercer County Community College in New Jersey used to boast a graduation rate of 87 percent of its enrollees, says Robb Smith, the director. Last year, a quarter of the students dropped out. He lays the blame squarely on unrealistic expectations fanned by TV shows.

Tracy Lee doesn't even watch *Six Feet Under*. "That's the first question people ask," she says. "Oh, my God. If I watched it, I'd think we were crazy, too." Her own expectations of a life in funeral service have been somewhat tamped by school; she knows it's "just not that interesting all the time." Still, she's got to manage other people's perceptions.

"You get some weird looks when you bring out your textbooks at Starbucks," she says. "The good thing is you always get a subway seat when you're carrying a book with the big fat title *Embalming*."

Embalming is the skill most people associate with mortuary school, but the curriculum demands a lot more than that. Most schools require forty-five credits toward graduation, a process that typically takes two years and results in an associate's degree (some states require four-year bachelor's degrees in the mortuary sciences), to be followed by a one- to three-year apprenticeship. Schools require classes in embalming, of course, but also anatomy and physiology and pathology, as well as "restorative arts"—in which students learn how to make a corpse presentable. There's business management and accounting and computers, not to mention law and ethics and grief counseling.

Even with the focus of funeral directing shifting to aftercare and personal service, and with the steady rise of cremation as the burial option of choice, embalming is still the subject most students—76 percent, according to the *American Funeral Director* poll—find most interesting. AAMI doesn't have a morgue of its own, so students hike across town to the New York City morgue. (Dunn rejects my requests to attend these classes, citing state law forbidding outsiders.)

There, they practice embalming cadavers that have remained unclaimed for at least six weeks. That period seems short to me, and in fact, Tracy says one time a body was claimed right off the table midway through an embalming class. The students aren't told how the people died or who they were or, indeed, anything at all about them.

So far this semester, Tracy has learned to insert eye caps, which are exactly what they sound like: large, contact-lens-like caps over which closed eyelids look most natural. She has learned how to shut a gaping mouth with a giant needle and thread. She has learned to always start with the feet, "so it's not so personal."

But one thing she hasn't learned yet is how to deal with the smell. "It smells—like death," says Tracy. "I mean, I can't describe it any other way. It's not formaldehyde. It's not any sort of chemical smell. It's death. And it lingers. I walk in the door and I shower."

⚫━

Somewhat less pungent but no less important is Funeral Service Principles, a course in the nuts and bolts of the business.

The first thing I notice when I arrive for my first class is that I am the only one wearing jeans. AAMI has a dress code in keeping with the industry's: ties for the gents, dress shoes for the ladies, everything in dark, conservative colors. The only one dressed less appropriately than me is the skeleton by the blackboard.

I make my way toward Tracy, who waves from the back. A student behind her is wrapped in a floor-length down coat. "It's cold as a morgue in here," Tracy hisses. Other than the funereal attire of the students, the classroom looks like any other with its fluorescent lights and orange floor tiles and ergonomically criminal chairs.

The class is taught by Father John Fraser, a funeral director turned Catholic priest. His pink face is anchored by a white priest's collar. He has a light voice and a theatrical delivery. In a different lecture I attend on Roman Catholic funeral rites, Father Fraser dresses up the skeleton in vestments and a pink skullcap. "Trouble is," he murmurs to the skeleton, "you look exactly like a lot of bishops I know."

He tells me later that he discovered his priestly calling while working as a funeral director, squiring priests around from church to cemetery. The career shift is not all that surprising: When you work with the dead, I bet you seek some answers. He probably wasn't thinking about this at the time, but imagine the shingle you could hang as a priest-slash-undertaker. Embalming and Eucharists. We Bless and Bury.

Father Fraser is discussing caskets today as part of a segment on funeral merchandise. He ranks materials by quality, listing various wood finishes—from mahogany to South Carolina poplar with mahogany finish to corrugated fiberboard with no pretense of mahogany anything. "So what? Who cares? Why are you torturing us with this?" Fraser asks himself. "For one reason: appearance. The service you're offering is to give the family a suitable memory." I think the message is that if a family can afford only plywood, you need to know how to make it look like mahogany.

Fraser goes on a while longer about laminates and stippling. It's beginning to sound like a Home Depot workshop. There's a buzzing sound directly in front of me, which after a while I recognize as snoring.

A student raises his hand. "Realistically," he says, "how often would we have to explain all this to a family?"

"Once in a lifetime," Fraser says cheerfully, "*maybe*. But it's six points on the boards."

The lesson, and mortuary school in general, can be as dry as corrugated fiberboard. In one sense, that's a realistic reflection of the industry: Funeral work isn't all glamour all the time. There are eye caps and airbrushed cosmetic reconstructions, and then there are state forms and jammed fax machines.

The point is that the national boards and licensing exams are full of snore-inducing questions about laminates and stippling, and Fraser wants his students to pass. In a lively voice, he suggests mnemonic devices, such as "BM" for the procedure in which metal caskets are "bonderized" with "manganese," a question on the exam that "could cause gastrointestinal distress."

Toilet humor from a priest in a lecture about caskets. I'm beginning to like mortuary school.

———

Funeral education didn't used to be this hard.

Until recently, virtually every young mortician was the son (occasionally the daughter) of a mortician, who was also the son (almost never the daughter) of a mortician. Undertaking was a set of skills you learned on the job, at your father's elbow. You attended a year of mortuary school, aced the piece-of-cake boards, and went back to work.

Then the industry came under scrutiny, thanks largely to one Jessica Mitford. And students with no family connection to the business increased from one in two in 1971 to two out of three in 2003. Standards would have to change.

Becoming a funeral director today is a four-step process that generally takes at least three years. Aspiring undertakers must attend about two years of mortuary school, pass the national boards, pass the state licensing exam (or exams, depending on whether your state requires a separate license for embalmers), and complete a one-year-minimum internship.

In 1996, after ten years of haggling and delays, the industry finally put some teeth in its national board exams, which are now administered electronically on computers at H&R Block outlets. "It was a long time in the making," says George Connick, executive director of the American Board of Funeral Service Education. Educators say the number of questions on the board have actually decreased by half—from six hundred to three hundred—but the subjects covered have widened and deepened, the questions become more pertinent. "It has changed to take into account changes in the funeral service itself, which are mostly cultural and scientific," says Connick.

For instance, the exams used to ask generic science questions like how many bones there were in a human body. "Well, an embalmer doesn't need to know that," says Smith of Mercer County Community College. But he or she does need to know which of the following sutures would best be employed for an abdominal puncture wound:

A. lock

B. baseball

C. intradermal

D. purse string

And in posing the eyes, the eyelids should meet at the:

A. center of the orbit.

B. superior third of the orbit.

C. inferior third of the orbit.

D. middle third of the orbit.

(Answers: D. and C.)

The tougher exams begat tougher curriculums, which now take an average of two years at most schools to complete. A degree from AAMI takes twelve to sixteen months total. What's more, as times and communities changed, schools also began offering lectures on the funeral rites of Hindus and Muslims as well as workshops run by grief counselors. The staff at AAMI includes a medical examiner for the city of New York, but also a certified family therapist, a lawyer, a nurse, an accountant, and a chiropractor.

Critics say it's not enough. "Funeral education is still way behind the curve," says Josh Slocum, executive director of the Funeral Consumers Alliance. He notes the paucity of cremation mentions in the curriculum and in the exams. "Mortuary school is not rocket science. It's not med school. It's a trade school. There's nothing wrong with that. But the candid student will tell you the courses and focus are still out of step with what consumers really want."

———

I meet Renaye Cuyler in AAMI's library, which houses an impressive collection of books with "death" in their titles. She is middle-aged, bespectacled, carrying a briefcase and could be mistaken for faculty. Renaye is in fact a practicing attorney who, at the age of fifty-six, is seeking a fresh career in mortuary services. In 2006, she and a partner hope to become the first black women in Brooklyn to open their own funeral home.

They surely won't be the last. Blacks have a long tradition in funeral service; the National Funeral Directors & Morticians Association, an organization of black funeral directors, was founded in 1923, and Renaye's own grandfather was in fact a mortician. But their ranks are growing. African Americans made up 29 percent of mortuary students in 2003, up from 12.6 percent in 1971. That's fascinating, considering blacks make up only 13.3 percent of the U.S. population. Hispanics and other minorities make up another 9 percent of mortuary classrooms, up from 2 percent three decades ago.

Funeral homes, like churches, remain segregated in America. Black families have always gone to black funeral homes, which in turn have always been independently owned. This lure of capital ownership is what has drawn Renaye Cuyler to funeral service. Sitting across from me at a table in the library, Renaye tells me that after thirty years of practicing law, she found herself contemplating a retirement portfolio nowhere near what a successful law career should have yielded.

"Personal injury law just isn't that profitable anymore," she says.

Meanwhile, her friend Maria Sealy, a funeral director at a home in Brooklyn, was contemplating striking out on her own. Neither Renaye nor

Maria comes from funeral families. Renaye's grandfather did once own his own business, the first black-owned funeral home in Williamsburg, Virginia. But when none of his ten children expressed interest, it slid out of the family's hands. Renaye carries his business card in her day planner as a good-luck token:

FOUNDED 1926 TEL20F5
GEORGE E. B. TABB & CO.
Undertakers
With Entire Motor Equipment
Day and Night Service

"He's my angel," she says, pointing a finger toward the ceiling.

Anyone starting up a funeral home could use divine oversight. Renaye and her partner began by writing out a business plan. The demographics argue the business potential, says Renaye. "Brooklyn is the rare community where death statistics continue to go up, due to diabetes, violence, HIV. It's gruesome," she says, and here she exhales a short, rueful laugh. "But somebody has to handle these bodies."

These grim statistics, spelled out right in the business plan, didn't exactly ring *cha-ching* to bankers' ears. Renaye contends funeral homes just aren't well understood in the banking community, though she entertains dark suspicions about other reasons for their frequent rejections.

In the end, a staffer at the Small Business Administration took up their cause and helped them locate a bank—in Utah, of all places—that offered up half the funds. The SBA put up another twenty percent, which left Renaye and Maria to come up with 30 percent on their own. The 12,380-square-foot lot cost $839,000. Added to the cost of construction, the bill will total $1.3 million. That puts the women's contribution at about $390,000. There went the retirement fund whose slender profile had put Renaye on this road to begin with. "We've mortgaged everything but our children," she says.

The week I meet Renaye, she and Maria had just closed on an empty lot in Prospect Heights, an area they had selected because it was not immediately served by a competitor. When I see the lot some months later, I see why. Even so, when I Google the address, I find there are twenty-six other funeral homes within a ten-mile radius.

———◈———

Despite the influx of funeral service newbies, for the time being there are still plenty of trainee morticians who won't ever have to worry about wooing a bank in Utah or plowing through SBA mumbo jumbo. At least a third of current mortuary students are in position to inherit a funeral home.

Lou Stellato III—"Junior" to the staff of the Ippolito-Stellato Funeral Home in Lyndhurst, New Jersey—is one of them. Though he'll never have to ask what a pall is or interview for an apprenticeship, Lou Jr., twenty-two, still has to complete his degree like everyone else and pass his boards. He's currently in his second year of studying mortuary sciences at Mercer County Community College.

Lou Jr. is a younger version of Big Lou, proprietor of the business and his dad: stocky, friendly, quick with a handshake. Every time his grandmother goes to the beauty parlor, she's besieged with the phone numbers of other people's granddaughters. He's got future mayor written all over him.

Just as women and minorities swarm the business, the likes of Lou Jr. are getting out. Growing up, they witnessed their parents ditching Disney vacations for last-minute services and endlessly Endusting the casket showroom. They want no part.

But Lou Jr. never really considered another life. He and his two sisters learned to roller-skate in the carpeted viewing rooms of his father's funeral home. He's done retrievals since he was fourteen. His high school pitching trophies crowd a glass case in the basement of Ippolito-Stellato, directly across from a shelf of books on grief. His bachelor's degree in biology from Drew University is framed on his father's office wall. He's lived his entire life on the second floor of the funeral home. If for nothing else, what other job could beat that commute?

Two days a week, as stipulated by Mercer County's program, Lou Jr. apprentices at his father's side. In New Jersey, aspiring undertakers must complete a one-year residency at a certified funeral home, during which they attend embalmings and services. Typically, those residencies resemble indentured servitude; you're the on-call guy for 2:00 a.m. body removals, the go-to girl for typing up a death notice for the *Star Ledger*. For that privilege, some earn maybe $20,000 a year; others earn nothing at all.

One frigid Tuesday in January, I shadow Lou Jr. on the job to witness the life of an apprentice undertaker.

By the time I show up in Lyndhurst at 9:00 a.m., Lou Jr. is dressed impeccably in a black suit and silk tie, shoes shined, hair gelled. He's standing in the foyer of the funeral home, slapping backs and pumping hands as he greets today's part-time staff, high school pals who help carry coffins and do other odd jobs. "How you doin', Mike? You lookin' good—I like that shirt."

At 9:15, he hops into the flower car, a sort of fancy flatbed used to transport the giant standing floral arrangements typical at Italian American funerals. He's whistling and coatless. The flowers have already arrived, so right now he's using the car to get to Holy Cross Cemetery to "clear the papers." This is a service Stellato is proud to offer, this preburial administrative clearance; otherwise, families have to wait in the cold while the office ladies shuffle forms.

Lou Jr. hands the unsmiling woman behind the counter a stack of paper. She scowls: He's missing one duplicate of one form. (I am constantly amazed at the paperwork generated by death care providers. Out of the stack, only one form is required by the state, says Lou Jr.; the rest are for the cemetery.) The other workers, matronly ladies with permed hair, smile sympathetically. Lou Jr. whips out his cell phone to fix the problem, then calls out hearty good-byes and thank-yous. I can see the ladies' eyes shining as they envision Lou Jr. walking down the aisle with their Jennifer or Deanna.

This morning at 10:00 a.m., he will run the funeral of an elderly local lady. About three dozen family members are already gathered in the larger viewing room. The lady is laid out in a beige-colored metal casket, her auburn hair expertly coiffed by Lou Jr.'s mother, Linda, and her sparkly, powder blue dress from the Stellatos' collection of burial wear (merchants sell clothing designed specifically to dress the deceased, with open backs for easy dressing and strands of fake pearls preattached). She looks peaceful and unreal.

Above her head looms a bleeding heart, a classic arrangement at Italian funerals that involves red roses in the shape of a heart with dangling red ribbons to symbolize the blood. One flower arrangement represents an open Bible, another a crucifix. Catholic symbols aren't cute.

In the back of the room, a TV is on. Images of the lady flash slowly—standing beside her husband in a faded photo, surrounded by a row of beaming grandchildren in another—interspersed with winter scenes and accompanied by tinkly Muzak. Before each service, Big Lou asks families to bring in photographs, which he then scans and sends out to a business in Portland, Oregon, that specializes in video tributes to the dead. So many funeral homes now offer these tributes that it's often included in the price list. Stellato throws it in as a surprise bonus.

"For Lyndhurst, this is cutting edge," Stellato explains later. "I've never had a family not love it once they see it, but suggesting it beforehand might freak someone out."

Lou Jr. suddenly realizes his dad is not around, and he will have to lead the prayers himself. This being a Catholic funeral, the religious service will take place in the church, but the viewing typically calls for brief prayers.

Lou Jr. steps over to the widower, who is hard of hearing, to let him know the prayers will now begin. The crowd watches him expectantly. Lou Jr. then laces his fingers together and recites the Lord's Prayer and then the Hail Mary. The guests follow. As they rise to file past the casket, he waits by the widower's side, then pushes the wheelchair up to the half-open casket.

The old man reaches out and grasps his wife's hand. "Good-bye, good-bye," he says, sobbing loudly. Lou Jr. stands at a respectful distance and waits. When the widower's son nods, Lou Jr. darts to open the door for the old man's exit.

Mortuary school teaches budding undertakers the science and business of death; the apprenticeship teaches them to anticipate. Watching Lou Jr. at work, I can best describe his job as the purveyor of small kindnesses. It's the tug he gives the widower's belt so as not to expose his backside as he eases the old man from his wheelchair into the limo. It's the wordless assistance two of his colleagues give a man with crutches in scaling the church steps. It's the blanket Lou Jr. fetches out of the limo's trunk at the snowy cemetery, just in case someone gets cold.

Only experience can teach you to anticipate a mourner's shiver. They don't teach that in school.

Back at the funeral home, the staff is rushing around, preparing for two more viewings. The rooms buzz with power vacuum cleaners (a tool so vital in the funeral business that the national convention had a booth sponsored by Oreck). Linda Stellato, a pretty blond woman whom I last saw in a chic black outfit at the holiday grief ceremony, is dressed today in sweatpants, yawning, using a curling iron to set the hair of an elderly female corpse. She looks up as her husband sets up a row of priests' garb, embroidered in gold and hanging on easels.

"Three vestments—huh," she says, eyebrow arched.

"*And* a pall," says Stellato, corralling Lou Jr. to fold the heavy embroidered cloth that covers caskets in church. Families buy these priests' accessories and then donate them to the church as an offering in the deceased's name; such an array of gifts signals this family's prominence and wealth.

Lou Jr. saunters over to the casket. "Hey, Mom, you oughtta take a look at the gentleman in the other room," he says casually.

I ask why.

"I did him," he says, beaming. "My first embalming."

We go to check him out. The gentleman is laid out in a lovely casket surrounded by flower arrangements. He has white hair, a prominent nose, and bright orange skin.

We stare at him for a while. "I think he looks okay," Junior says, sounding doubtful. "But all I hear is criticism."

I ask what's wrong.

"*I* don't know. *He* says he's not positioned right."

He walks past. Big Lou pauses by the body and purses his lips. "Their faces, they ought to be angled at five to fifteen degrees," he says. "So that they face you a little bit instead of being in profile."

Indeed, the gentleman's nose points straight at the ceiling, giving him a haughty appearance. His father is right. Lou Jr. sighs.

These details will take time to learn. But Lou Jr. is entering the business at a time when such things may not matter much longer. Lyndhurst's 15 percent cremation rate is low compared with the 25 percent nationwide, but even so it's up from none at all a decade ago. At Ippolito-Stellato, even the families who cremate choose to embalm their dead first and to hold a traditional viewing before the cremation. Still, viewings that once lasted two to three days are now in many cases reduced to a few hours on the day of the service. Only the family turns up instead of the entire town.

So if a handful of folks will look at the body for just an hour or two, why work so hard to become an expert embalmer? As more people begin to refuse the procedure altogether, can embalming help but become a dying art?

Mercer County Community College is a commuter school, a collection of virtually windowless cement-block buildings, proud remnants of 1960s-era New Jersey architecture. Plopped in farmlands near Trenton, the school draws students from Pennsylvania and New Jersey, from recent high school grads to midlife career changers.

They gather in a small lounge, making bleary chitchat over lukewarm coffee before classes start at 9:00 a.m. There is no dress code at Mercer as there is at AAMI, so they're slouched in oversize sweatshirts and baseball caps. In appearance, they differ not at all from the accounting students down the hall, though their conversation sets them apart.

"When you place a flag on a half-open coffin, how many times do you fold it back?"

"Three."

"How about the pall—do you face the cross toward the family or toward the crucifix?"

"The family, I think."

"Naw, it's the crucifix. I'm pretty sure."

The Christian rites are complicated enough, the students grouse, never mind the unfamiliar religions that are beginning to infiltrate their communities. A young woman tells about a Muslim service conducted recently by the Princeton funeral home where she serves as an apprentice. "I had to hide out in the van," she says. "All the other women were wearing veils, and I couldn't be seen."

The school is run by Robb Smith, a second-generation funeral home director from south Jersey and a dead ringer for former Republican presidential candidate Lamarr Alexander. He keeps an urn labeled "Ashes of Problem Students" front and center on the messy desk in his office. (It's a variation on a similar urn serving as the tip jar at a coffeeshop in Nederland, Colorado, site of the Frozen Dead Guy Days festival, which read, "Ashes of Problem Customers." You have to take your funeral humor black.)

Smith has practiced for over thirty years and taught for nearly as long. I am curious to witness funeral education as filtered through a current practitioner.

Smith writes the word *protective* on the whiteboard. "Protective" is a word some funeral directors, despite two-decade-old rules prohibiting them from doing so, still use to describe caskets that seal—as if sealing protects a body from decaying. In fact, sealing a casket can interfere with the natural decomposing process, causing bodies to fill with gases and eventually blow up. Amazingly, the word still appears on the national boards to describe a certain type of casket.

Instead of glossing over the inaccuracy or, worse, reinforcing it, Smith points it out. "Remember who writes the test," he says. "There are those funeral directors out there who haven't cleaned up their act."

Despite the dramatic revisions made to the tests in 1996, the boards and, thus, funeral education still reflect old values and practices. In Smith's office is a blue plastic binder that holds this year's syllabus, in which big chunks are dedicated to embalming and communicable diseases. But there's no section on cremation.

Let me repeat that: In a country where 25 percent of its dead were cremated in 2003 and 48 percent will be in 2025, mortuary students do not take a course on cremation.

Smith says that will eventually change. For now, cremation works its way into other subjects, such as rules and regulations of funerals (how long must one wait, by law, before a body is burned?), counseling (does a family choosing cremation experience grief differently?), and merchandise (is it wise to keep an urn showroom?).

Cremation comes up in today's lesson, which poses the question of what a funeral home ought to do with cremains that are left unclaimed. This happens alarmingly often, producing a headache for funeral homes whose storerooms fill up with the small, heavy plastic boxes. The International Cemetery and Funeral Association advises that the next of kin be notified by registered mail after 180 days, then given another 90 days to collect the remains. In New Jersey, the law requires that the home wait a year and make a proven effort to contact the next of kin. After that, the ashes can be disposed of.

But how? Where? "You guys are too young to remember *Hill Street Blues*," says Smith. "When this one cop dies, they sprinkle his remains at a certain intersection in the middle of the road. Then in the next scene, you see a street sweeper come by."

For a moment, Smith has a dreamy look in his eyes that makes him look even more like a presidential hopeful. Then he comes back. That kind of Zen symbolism—who are we, in the end, but dust on a street?—might work in Steven Bochco's universe. But in real life, ash disposal by public sanitation probably isn't the wisest option for a funeral director hoping to retain his job.

One student suggests buying a crypt in a mausoleum and keeping unclaimed ashes there. Smith nods.

Then another student asks: "If a family member eventually comes to ask for the remains, can you charge them a grave-opening fee?"

I'm floored. It's customary for funeral homes or cemeteries to charge a fee to reopen a grave; after all, a lot of work will go into digging the hole and hauling out a casket, though if it's enough to warrant a four-figure bill, I don't know.

But charging to open a door and fetch a box of ashes you're dying to get rid of anyway? That coldhearted focus on money surprises me, coming from what I thought would be dewy-eyed future undertakers. They're months from graduation, and already they're crossing the intersection of death and business.

Smith considers the question. "It's not in the rules," he says carefully. "To cover yourself, I'd suggest telling families in a letter. Let them know you'll charge a fee to collect ashes unclaimed after a certain period."

I shouldn't be shocked at any of this. Death is a business, and students are here in part to learn how to profit from it. Indeed, business, the subject, takes up a thick slab of the syllabus and makes up half of the boards. Despite the scourge of consolidation by big funeral chains, 89 percent of funeral homes are still independently owned, according to the National Funeral Directors Association—meaning many of today's graduates need to be able to discuss payrolls and inventory, have the hang of turnover analysis and trade discounts, know their EOQs from their CODs and MOMs.

Balancing the books is one thing. The millennial mortician must also know how to hold a hand. The psychology of grief is a subject long explored by mortuary students, but today's programs supplement the textbooks with touchy-feely workshops. At Mercer, grief counseling is taught by a couple, Deb and Neill Tolboom.

Deb Tolboom is a licensed counselor, funeral director, and former nurse; her husband, Neill, is a United Methodist minister and former technology officer at the brokerage firm TD Waterhouse. He says he's a doctoral candidate in funeral services, an academic accomplishment I didn't even know existed. They're energetic and goofy and lob throwaway questions on quizzes ("Are burdensome homework and lonely boy/girlfriends and looming exams stressing you out? Yes/No"). They finish each other's thoughts, which can be a little distracting midlecture. They even look alike: Both are sort of round and smiley and huggable, like Hobbits of a slightly gloomy shire.

The Tolbooms are of the mind that any celebration of a life lived, no matter how unconventional, is worthy. "We recently spoke to a couple who had lost their eighteen-year-old daughter," says Deb. "She used to like to drink. Well, her friends were gathering at her grave and leaving beer cans."

Neill nods. "They thought it was awful," he says. "They hated it."

"But we say, hey, that's great—"

In unison:

"They're remembering her—"

Deb finishes, "So she's not forgotten."

Grief counseling—or aftercare, as it's called—is a growing movement among some funeral directors. Their efforts go beyond offering a shelf of brochures featuring soft-focus photos of sad models to real treatment for real distress. Carmon Funeral Home in Windsor, Connecticut, founded and partners with a "grieving center" for children and families called Mary's Place. Renaye Cuyler and her partner, Maria Sealy, plan to build a counseling room right into their new funeral home, complete with a staff counselor.

These professionals feel the business of arranging funerals is a perfect opportunity to aid those in their deepest grief. Some, like the Tolbooms, feel it is an obligation.

The students aren't convinced. In an earlier class, the Tolbooms had invited guest speakers who strongly urged having a grief counselor present when a family arrives to make arrangements after a death. Some in the class balked at that, saying the counselor's uninvited presence badly oversteps and intrudes upon a family's privacy. Clearly, grief counseling has some image problems among the undertaking set.

Today, the Tolbooms have asked the students to act out the dos and don'ts of grief counseling. Joe, a big lug of a student in a blue polo shirt and shorts, plays the counselor; Jen, a slender blonde wearing stately jewelry and a black suit, plays a recently widowed woman. The widow, whose husband was twenty years her senior, has been struggling to cope.

SCENE ONE: Don't. Jen comes through the classroom door wearing big sunglasses. Joe throws his arm around her, escorts her to a chair, then drags his own chair around the table right next to hers.

JOE: Hope you don't mind my outfit. I was drinking beer before, and it spilled.

JEN: I . . . can't seem to get out of bed. I just can't function. Every time I go in the room where it happened, I just . . .

JOE: So I guess this means you're single now?

JEN: What? Oh—what?

JOE: You wanna go out for a drink after this?

JEN: I don't drink.

JOE (helpfully): Drinking can help in situations like this. Drinking, drugs. You gotta get out there, you know. He ain't coming back.

Neill Tolboom holds up a hand to stop the laughter in the classroom. "What did he do right?" he calls out. "He opened the door for her—good. Yes, that was about it."

The two prepare for the "do" scene. "Now let's see you act," a classmate calls out.

SCENE TWO: Do. Joe meets Jen at the door with hand extended.

JOE: I'm sorry for your loss, but hopefully what we talk about today will help you get through your grief. (He offers her a seat and then a drink of water, folds his hands, and listens.)

JEN: I lie in bed all day. We had a dog. I mean, we still have it. I have it, I guess. Anyway, I forget to take the dog out. (She dabs at her eyes with a tissue.) I'm so worried. I don't know anything about money— he always handled that. I guess there are like stocks and bonds, but I don't even know what to do about them. I can't take care of the house. Mostly I just feel so sad. Like, is this normal?

JOE: You should know you are simply going through the steps. This *is* normal. Over time, we will work through this together.

This time, Neill is full of praise. "First, you acknowledged her grief. Then you made her comfortable. As a funeral director, those two small things are like the most important things you can do for someone in grief. You told her she was normal. Most people in grief feel like they are the only ones who have ever felt this way. Most important, though, you gave her hope. You told her there was a way out."

Deb steps in. "Look, we don't expect you to be counselors. But you absolutely owe it to your profession to help these people. You can be the purveyors of hope."

Renaye Cuyler gradated from AAMI in August 2005. In October, I drive out to Brooklyn to see the site of Renaye's funeral home. Construction crews had broken ground in March, and building was well under way; they hope to open the following spring.

I travel down Atlantic Avenue past the antique shops and Asian fusion restaurants, past the bodegas and the check-cashing storefronts, past the spot welder and Leo's Bent & Cracked Rims. I inch past a tow truck hauling a red Mazda with a spiderweb crack in its windshield and pull up behind a silver SUV with its side sheared off. There, on a tall metal fence about twenty feet long, is a sign:

FUTURE FUNERAL HOME

I get out of the car and wait. The sky is brilliantly blue, the kind of day on which it's hard to feel threatened even in a neighborhood I would describe kindly as industrial, less kindly (but still kindly) as dodgy. Next to the metal fence is a boarded-up town house, on whose door someone has scrawled "DANGER—KEEP OUT—OR ELSE!" The S train rumbles by on an elevated track. A transvestite in an orange shirt saunters past, stopping to chat with a man pushing a shopping cart full of tire rims. A clutch of men sit on folding chairs at the end of the fence, staring and laughing at the trannie. Or maybe they're laughing at me.

Renaye and Maria walk up. They look for all the world like a pair of no-nonsense grandmothers out for their daily constitutional, which in one sense is what they are—albeit grandmothers neck deep in a $1.3 million gamble.

Renaye Cuyler and Maria Sealy have known each other for sixteen years, and they look almost exactly alike. Both are fifty-six. Both have close-cropped hair and glasses and sneakers and matching cell phones clipped to their belts. Renaye is Baptist and Maria Catholic, but both wear large crosses around their necks.

We step through the gate and onto the property. Inside the metal fence, a huge pit has been dug into the ground; at the last minute, they had decided to add a basement. Half a dozen men speaking Spanish and Chinese and English look up. The site manager waves.

"How's it going?" Maria calls.

"All right, all right," he says, wiping his brow.

The property was formerly an empty lot, of which there are a surprising number in this part of Brooklyn. Renaye and Maria, too, were surprised, given that they live six blocks away where one-bedroom apartments are going for $800,000. They have firm hopes for this neighborhood, which according to them will transform from industrial/dodgy to gentrified/up-scale within the decade. For now, funeral guests will just have to wend their way around the auto graveyard. The pair hope to buy the bombed-out town house as a catering hall in which to offer postfuneral repasts, an important part of the African American death ritual, but for now the $390,000 asking price puts it out of reach.

Renaye pulls an architect's rendering out of her bag. "You see, we'll have five chapels, or viewing rooms," she says. "A prep room on the first floor. And—this is key—seventeen-car parking out back."

"That's key," Maria echoes.

Maria is from the Bahamas. She ran a chain of liquor stores there until she decided to attend mortuary school in Miami Beach. She opened a funeral home in Nassau but left the business to her son to pursue opportunities here in New York. "People thought I was crazy when I left the liquor store business," she says. "It was very lucrative. *I* thought I was crazy. I was scared of dead bodies."

When I ask her what made her change her mind, she hedges. "I don't know if I want this in there," she says.

Finally she relents. "Well, I had a vision. I was driving around my friend Paula's neighborhood when I saw something I'd never seen before. It was a white funeral home, with a white hearse parked in front. I told my friend Paula, and she says, 'Maria, there's no funeral home there,' she says. I drove back there again and again, and I never saw it."

I tell Maria that a vision is as solid an indicator as any that she made the right choice. When the choice you're making is to become a funeral director, a vision that involves white hearses holds more weight in my view than, say, a Myers Briggs career test.

According to the American Board of Funeral Service Education's Web site, motivations to enter the profession include a "desire to be of assistance and to work in a human service profession. . . . Others view funeral service as relatively 'depression proof.' "

Although the mortuary students I met were indeed remarkably cheerful, well-adjusted individuals, I think here the board meant "Depression-proof." It's true millennial morticians-to-be look nothing like their predecessors and face a far different landscape, too. But whatever their reasons for choosing this profession, one thing's for sure: The board is right. In this world, nothing is certain but death and taxes. Like accountants, they'll always be sure of a living.

Orchids and Chopsticks

FUNERAL RITES IN THE OLD COUNTRY

*Opapa liked cars. He would have enjoyed his last ride
in a dragon-led chariot.*

The call came in the night. "Opapa has died," says my mother.

My Japanese grandfather—Opapa to us—had taken a turn
for the worse just as my mother had arrived in New Jersey for the Christmas
holidays. She had immediately boarded a plane back to Japan, leaving gifts
color-coded for their intended recipients, listening over the phone from the
hospital to our children's squeals as we opened them on Christmas Eve.

Your beloved relative has died. Like so many Americans, I hear these
words over the slight static of an international phone line. Born overseas

like 12 percent of the U.S. population, I am bound to immediate kin many thousands of miles away. I had in fact hoped over the course of reporting this book to explore the effects of foreign roots on immigrants' death traditions. What rites did they bring with them from the old country, and why? I just had not predicted the immigrant in question would be me.

The Japanese, much like Americans, are currently reinventing their own funeral rites. Nine out of ten funerals conducted in Japan involve Buddhist ceremonies, which in turn involve complicated ritual and shocking expense. The average funeral in Japan costs $32,000, according to a 1995 government survey. Imagine the cash Japan's funeral industry stands to collect as the current annual death rate of one million soars 50 percent by 2020.

My hope is that, by then, more Japanese will rebel and demand what they want. There are already signs of revolution. A young wedding planner in Tokyo quit his job to start a "funeral production" business, according to newspaper and magazine profiles. Another new company called Ending Plan offers intimate family funerals at a low cost. A popular actress shocked the nation by conducting her own funeral while still alive, then later having her ashes scattered in the bay. Hotels are trying to make up for dwindling bookings for weddings by offering packaged funeral banquets. Never a land to miss out on weird merchandising opportunities, Japan is also first in the world, I venture to guess, to market eyeglasses designed specifically to wear during funerals.

Cremation being the disposition of choice in virtually every death, the Japanese are beginning to hunt for creative things to do with ashes. LifeGem, the Chicago company that markets diamonds made from cremated remains, launched its Web site in Japan in 2004 and received one hundred thousand hits the first day. I read a newspaper story of an entrepreneur who offers bereaved pet owners the dignity of a cremation with the convenience of door-to-door service: His mobile crematorium will incinerate Taro at the curb.

Space is precious in this country the size of California with a population half that of the United States. The dead, like the living, are being squeezed. There's a nine-story columbarium in a crowded neighborhood of Tokyo. And then there's the virtual cemetery. Here's a news item from the Web site of the Ministry of Foreign Affairs:

A temple in Hiroshima opened a cybergraveyard this past April that allows users to choose the type and location of the burial site free of charge. By accessing the temple's web site and registering the name, profile, and photograph of the deceased, visitors are presented with an engraved tombstone created with computer graphics. One can even sign the register of guests and offer flowers and incense by using the keyboard and mouse.

In Tokyo, I meet with John Kamm, an American whose Colorado funeral business goes three generations back. While studying for an MBA at the prestigious Waseda University here, he learned of Japan's notoriously over-priced, corrupt, and unregulated funeral industry.

"Every transaction involves a kickback," says Kamm, who is thirty-four. He compares the industry with that of the United States before the Federal Trade Commission implemented its funeral regulations in the 1980s. What's more, he adds, "the people are too polite and embarrassed to bargain or to question the costs." Kamm launched All Nations Society, a funeral co-op much like the Neptune Societies in the United States, an organization with members who prepay for affordable services.

I am always glad to hear of change in my home country, particularly those benefiting consumers. But for my eighty-seven-year-old grandfather, a practicing Buddhist and prominent member of society, I know we will hew to the ancient rites of death.

TOSHIAKI TAKEUCHI
1917–2004

*T*raveling to funerals has become a broadly American rite as our families have scattered far from our hometowns. Traveling to funerals overseas on no notice is even less fun. Traveling to funerals overseas on no notice with an infant is . . . well, you might imagine. My sister and I plead with various authorities to obtain speeded-up passports for our babies. My brother leaves a heavily pregnant wife in London. My other brother abandons his family on their Vancouver ski trip.

I don't sleep much during the twenty-four-hour trip, though, thankfully, the baby does. I know we can use the rest for what lies ahead. Here's what I know about the rites to come: They will be Buddhist, complicated, and foreign to my Americanized branch of the family.

I also know my own duties as the eldest granddaughter will begin this very night.

After my grandfather died, the family brought his body home to rest. He lies on bags of dry ice in a plain cedar coffin on tatami mats in my grandmother's house. There his body will stay until the service, and an immediate relative must keep him company—day and night.

The night we arrive, it will be my turn. My still-nursing daughter must join me, of course. Mika's first night in her ancestral land will be spent sleeping next to her great-granddaddy's corpse.

The first night after they brought him home, my mother, Opapa's eldest and only daughter, spent it by his side. Mama is devastated by the loss. Opapa was her guardian, her protector—the one who blessed her marriage to the American when others would not, the one whose guidance she actually followed.

His face, she tells us as we drive home from the airport, changed hour by hour during that first night. As she watched, the visage she had so recently seen twisted by disease and pain eased as the muscles gave in to gravity. The half-moon dumplings under his eyes, the grooves in his forehead, the trials and triumps of eighty-seven years—it all seemed to melt away. Even his dentures seemed to settle in.

All night, she stroked his face and talked to him. When they brought him in, the back of his neck was warm, Mama says. She wanted to throw aside the bags of dry ice, to hug his body close to her, to keep it warm if only for another day.

By morning, his neck was cool.

Opapa's body lies in the formal receiving room of my grandparents' house, a long, straw-matted space bare of furniture but for a pair of fancy altars honoring our ancestors. It was a great room for Chinese jump rope and wrestling. On New Year's Day, back when Opapa ran the family business, the room would be laid out with low tables from end to end with platters of sashimi and crab legs and rice cakes and giant bottles of our family's brand of sake. Hundreds of people—clients, employees, neighbors—would drop in to accept a shot or three. My sister and I would kneel stiffly in rib-crushing kimonos and try not to pour the hot brew in some dignitary's lap. Once, on a New Year's Day when I was fifteen, Opapa offered my hand in marriage to a red-faced young executive. He was completely kidding, but the incident mortified me for months.

"Opapa, they're here," my mother says softly as we kneel in front of his coffin.

The windowless room is frosty from the dry ice they have packed around his body. Across the length of the coffin is draped a silk cloth much like the pall used in Catholic funerals, but this one is embroidered with golden dragons. Bunches of incense and hulking bouquets of white lilies overwhelm my tired senses.

There is a narrow strip of balsawood bearing the Buddhist name, called the *kaimyo,* administered to Opapa at his death; each character costs $1,000 and is selected with much to-do by the monks, though I have read that many employ a software program that does the work for them. Because of the cost, the longer the name, the higher the status of the deceased. Opapa's Buddhist death name has eight characters.

Family members have set up a table of offerings. There is a big bottle of the family sake. It's opened and half-empty, the victim of a midnight raid by my drunken cousins on New Year's Eve. There are sweet rice cakes, a gluey bite of which would have laid waste to Opapa's dentures. And there is an open pack of Marlboros with a single cigarette pulled halfway out—for the man with cancerous lungs and a wife who harangued him at every puff. Apparently it's okay to enable someone who's dead.

On the lid of the coffin is a purple tassel attached to a small door. Mama gives it a pull. Suddenly, I am staring at Opapa's face.

"Everybody has come," my mother tells her father. She begins to weep. "See? They are all here for you."

I do not recognize my own grandfather. In Japan, dry ice is favored over embalming to achieve temporary preservation. But apparently the two methods have a similar aesthetic effect: This face appears plastic and wrinkle-free, the skin a waxy yellow, the mouth a straight line across the jaw like the Nutcracker in the ballet. A white silk shroud is pulled up to his neck, so that his head appears dismembered. Only his hair—still thick, still silver—seems real.

"He looks . . . younger," I say to my mother, patting her on the shoulder.

I do not know why we feel compelled to compliment the appearance of the dead. I have noticed this behavior at American funerals, too; someone always needs to mention how good the body looks. Why? After all, the dead absolutely and at last don't care.

"Doesn't he?" My mother nods.

"So young and healthy," adds my aunt. "And peaceful."

"Here," says my mother, handing me a Q-tip. I dab it in water, as she had, and touch it to his lips. To keep from drying out, she says.

In the far corner of the room are a pile of futon mats. We will sleep here, on these, tonight. We lay the mats out side by side with the coffin and pile them with down comforters. To keep the corpse fresh, the room has literally been turned into a freezer.

My brother Ken has grudgingly agreed to accompany us, though he remains reasonably convinced of the presence of Opapa's *yurei,* or ghost. Ken is a six-foot-two, 230-pound baby. He insists Mika and I sleep on the mat closer to the body. We douse the incense, and the house settles into silence.

"Liiii-saaaa. Liiii-saaa. I am Opapa's *yuuuureeeei.*"

"Shut up, Kenny."

"It wasn't me!"

"You're an idiot. It's frickin' freezing in here."

"Good night, Opapa."

"Gooooood niiiight, Ke-heh-heh-hen."

"Shut up, Lisa. Don't steal my joke."

The next morning, a crew from the funeral home arrives to cart away the coffin. The family members gather in the foyer. We bow from the waist as they carry the coffin on their shoulders out of the house. Our uncle leads the procession, carrying Opapa's *kaimyo;* the other uncle carries his black-beribboned portrait. His daughter scolds him for holding it with one hand instead of two.

In the few days before the trip, Mama had called repeatedly with instructions. "Black," she said. "You wear all black. No gray. No white. Okay? Nice black. Do you have?"

"Of course I have," I'd said, mildly insulted that she'd think a New York working girl wouldn't have a nice all-black outfit.

It won't do. When I dress the morning of the service in a silk sweater and a long skirt, my mother convulses with rage. "Nonononono," she barks, ordering me out on an emergency shopping excursion to the city's lone department store, where stock is limited to what in America we'd size a zero, petite.

"She's out of her head," I mutter to my sister.

"Totally," says Emy. Mama has ordered her all the way back to the house because she has forgotten to put on her pearls.

Mama's obsession with detail extends toward spiritual and traditional minutiae. This is a startling change in a woman who flouted generations of ancestral expectation by marrying an American and converting to Catholicism. As I sit the morning of the service trimming Mika's nails, she flies at us with a tissue, gathering up the microscopic nail clippings. They would be placed in Opapa's coffin, she says, to serve as currency in the afterlife.

"Do you want my nails, too?" asks Dad.

We all agree his crusty yellowed talons would probably incur debt.

The funeral will be held at an event hall, a cavernous, cement-walled space that can hold thousands of people. It has been decorated for Opapa's service with a tsunami of white orchids and chrysanthemums sweeping across the head of the room. At its crest hovers a miniature wooden temple. To one side of the temple is a pyramid of $100 melons and to the other a poster-size portrait of my grandfather.

At the foot of the floral tsunami lies Opapa's coffin. It looks small and lonely before this display.

Family members are to sit on one side of the room. Facing us across the aisle will sit the VIPs. The CEO of Kirin beer has come, says my uncle, who runs the family business now. And the head of UCC Coffee. And Ajinomoto. And 7-Eleven. They all resemble my grandfather in carriage and dress and waxed hair. But in no one do I detect his kind eyes.

Three monks file to their stations before Opapa's casket. By the time they begin to chant, the line of mourners snakes around the city block.

Funeral guests follow a tight script. Some wear suits, others wear kimonos, but every single person is dressed head to toe in black. Each makes an initial stop in the lobby. There they stand in line to hand over an envelope of cash and sign a register. Then they file up a ramp into the main hall, where they bow twice: once, deeply and at length, toward Opapa; once toward us, the relatives.

Each takes a pinch of pulverized sandalwood from a green marble bowl and transfers it to another bowl. Some pinch more than once, despite a stern sign decreeing one pinch per mourner. Putting their hands together, they repeat the bows and then move on down the line, where a funeral worker hands them a wet cloth. After wiping their hands, they shuffle toward the exit to receive a goodie bag containing a bottle of our family's top-grade sake. If they like, they can stop in one of the heated tents outside for a cigarette and watch the proceedings on closed-circuit TV.

I am seated in the baby ghetto of the relatives' section. Mika is getting antsy from the chanting and the incense, and I am running out of tricks to keep her quiet. Over the past few weeks, she has started to test-drive her language skills, trying out new consonant-vowel combinations. She is clearly preparing to check out the echo in the hall.

Just as the governor of Kochi prefecture takes the podium to deliver his eulogy, Mika lets loose with a brand-new sound.

"Blah blah blah."

"Shhhh!" I hiss.

My daughter smiles at me, delighted. *"Blah blah blah blah."*

I hustle outside, thanking Buddha that "blah" does not translate. Still, the governor looks displeased. Though he lives across the street from Opapa's house, they were not close; he is the brother of a recent prime minister, and local obligations like this one probably have him jonesing for the sophistication of Tokyo. It perhaps does not help when my sister later runs over his foot with her baby stroller.

The line of mourners is beginning to fray. I had stopped counting at about one hundred. I learn later that the final number tops two thousand.

—————

As the hall empties, the family is invited to gather around the coffin as its lid is removed.

His custom-made, three-piece, navy Yves Saint Laurent suit has been laid on top of Opapa's shrouded body. I can see his initials, TT, stitched in the collar. Each of us takes turns dabbing water on his lips. By the time it's my turn, Opapa looks as if he's drooling. My sniffle converts unexpectedly into a snort. Mistaking it for a sob, Mama hands me her hankie and hugs me. Now I quake with guilt.

It is time for him to go. The funeral workers bring out trays of orchid and lily and chrysanthemum buds to place on his body. At this point, my mother and aunts begin to wail. "Father, Father," Mama cries, tucking flowers tenderly around his shoulders. My uncle—the blank one who never says much of anything—stands at the head of the coffin and repeatedly strokes Opapa's forehead.

When we have buried him in flowers, the funeral workers fit the lid back on the coffin. They place slender gold nails in prescrewed holes and then hold out a golden hammer: Family members are to pound them in.

My mother waves her hands and backs away. "I can't, I can't," she says. I take the hammer instead. We take turns tapping down the golden nails until Opapa's box is shut tight.

The hearse awaits. Opapa would have dug these wheels. He loved cars, though he had a full-time chauffeur and couldn't drive himself for beans. He bought a new model every two years, often passing his barely used vehicles down to my family. Once he insisted on driving me and my husband on a tour around the city, and while doing thirty near an intersection he swung open his door to show us how it would trigger a recorded voice warning of impending danger. (Alas, I could not hear it over my own scream.)

Opapa's last ride would be a lavish one, protected in its journey by fearsome golden dragons atop the carriage. The funeral party follows, and the remaining guests bow deeply as we pull away.

In the United States, cremation often occurs unattended by family, in the inhospitable and industrial setting of a crematorium attached to a cemetery. This is something I think will change as more families choose this mode of disposition and insist upon a role. Immigrant groups are already doing so. At the Carmon Funeral Home in Windsor, Connecticut, I observed a Hindu funeral after which the family accompanied their loved one from the viewing room right into the adjacent crematorium.

In Japan as in India, cremation is a ceremony unto itself. This crematorium is a low, modern-looking building on grounds swaying with bamboo and burbling with fountains. We pull into the circular driveway and are ushered into a refrigerated chamber where Opapa's coffin already awaits. The monks—changed out of their purple-and-gold finery into everyday black-and-white robes—chant a final sutra.

The box is wheeled out of the chamber and into what looks like an elevator bank. We stop before door number seven. It opens slowly from the bottom, like a garage door. The uniformed workers push the coffin inside, where it continues to slide back through another door into the crematory. We bow again for what seems a very long time. I try not to think about *Soylent Green.*

Then we wait. The crematorium has a large waiting area for families, with a flock of servers to supply coffee and finger sandwiches. A cousin dashes out to buy *onigiri* rice balls and Häagen-Dazs. Two hours later on the dot, a voice over the PA system announces that our ashes are ready. The family gathers in yet another chamber, this one warm from the heat of the furnace.

A worker hands each of us a pair of giant chopsticks. Then we approach the table.

Cremation leaves a skeleton surprisingly intact. As I stare at the dusty bones, it hits me that this was a human being I knew and loved.

"There's his head," whispers a cousin.

"And his vertebrae," I say.

"What's that metal piece?"

"Oh, dear God . . . his dentures."

We begin at the feet. This is so the body won't be upside down in the urn, says my aunt. A tall, skinny crematorium worker wearing a pilled sweater over his uniform explains the procedures. My mother is elected to pick up the first shard, ceremoniously placing it with her chopsticks in the green ceramic urn. The rest of us follow.

I select what looks like part of a femur, mainly because it's big enough that I think I may not drop it. It's light and crushes like an eggshell in the urn as the worker packs down the pieces.

After they each take a turn, most of the relatives retreat. Soon only a few of Opapa's direct descendants remain at the table.

We work silently, picking bigger pieces first, moving from the legs to the pelvis to the back. There are bits that are reddish and blackish; when we ask, the crematorium worker says the color probably came from the suit and shroud.

He points out the Adam's apple, which everyone stops to admire. I'm told later the shape of the bone is supposed to resemble that of the Buddha. I wonder what Eve would say to that.

When we get to the head, the worker stops us again. The top of the skull is largely intact: round, smooth, rather small. He asks our permission, then quickly breaks it apart into pieces.

Once all the salvageable remnants have been packed into the urn, the worker seals it, places it carefully in a silver embroidered box, and wraps it in a gauzy white cloth. He points out which side was to face forward. My uncle takes the box, and we parade slowly out of the crematorium.

Back at the now-empty funeral home, we congregate in a side room where there is a photo montage of Opapa's life on the walls, blown up and arranged on crisscrossing stalks of bamboo. Here is the broad-shouldered young man squatting in a judo uniform. Here is the bespectacled soldier in his officer's uniform, about to depart for war.

Here is the proud father, hands on splayed knees, surrounded by three lively-eyed children and a breathtakingly beautiful wife. Here is the steward of a vast company his own father built up from a corner grocery. Here is the grandfather of eleven, all of us grown, none of us in jail. Here, the great-grandfather of thirteen, all healthy, though some blond and blue eyed.

Someone has ordered boxed meals for everyone. We sit at tables in that side room, exhausted, still in our funeral finery. Because we are a food family, and the death of the patriarch has not changed that, the *obento* is an elaborate affair involving sashimi and complicated vegetables. Because we are Japanese, and a death has not changed that, we take photos.

My cousins and I flank our grandmother, whom we call Omama, for the group shot. She is still beautiful, Omama, with her porcelain skin and round, bright eyes, wearing her black kimono embossed with the family crest. Seeing her seated at the center of our group, holding a portrait of Opapa, reminds me that they did very little apart. They traveled to Burma and Switzerland and Australia together. They ate together. They prayed together. She fussed over him. He indulged her.

The photos, when we share our prints later, turn out blurred and dark. I can just make out Opapa in his portrait, held tight by his wife, surrounded by his family.

Last Stop

SOME THOUGHTS AT THE END OF THE TOUR

I embarked on this journey cushioned against a fear of death by my distance from it. I am, after all, born in 1971 and in all actuarial probability likely to live another five decades.

Besides, I had mapped out a sunny road for my tour, one leading toward celebrations and funny trends and sparkly merchandise. I would skirt around that sticky swamp of religion. I would only occasionally encounter true grief, which would touch down like mist and quickly dissipate. The sad stories would all have happy endings, for who doesn't smile at a party?

But relative youth is no bulwark against death, as I learned. During the course of reporting this book, I lost two family members: my grandfather in Japan and an American cousin a week older than me. My families on both sides of the world are large but tight-knit, and the losses crumpled us.

Then, in November 2005, my mother was diagnosed with stage four cancer. Cancer is a known enemy in my family; she has battled it twice before, and won. Still, this latest assassin attacked with blinding speed and terrible fury, and for a time seemed unstoppable.

A month later, in Japan, I sat by my mother's hospital bed in the intensive care unit, waiting for her to come to after eight hours of surgery. I squeezed her hand and watched those green lines on the monitor hiccup. I would not think of death. To think of death signaled defeat. When an uncle pulled my brother aside to assure him all funeral expenses would be covered, we reacted with startled anger.

I began this book an unabashed advocate of the new American way of death, a way I believed involved celebration in place of mourning. Now I questioned my blithe conviction. If she died, if I lost this woman who raised me, would I have it in me to throw a party?

The sunny route I had mapped could not get me around the fact that death is a big, huge bummer. Losing someone, learning someone you loved is gone from your world, is horribly, crushingly sad.

Yet.

Death is also a release. It is so hard to say good-bye; it is much easier to remember.

On the day I learned the extent of my mother's cancer, I flew from my home in New Jersey to Washington, D.C., to cover the funeral of another family's beloved matriarch. It was a party, yes, with empanadas and dancing and memories and laughter. But before that there was a full funeral mass, at which the woman's daughters and granddaughters stood in the pews and simply sobbed. They did not deny themselves their grief, but they chose not to deny themselves a celebration, either.

It occurs to me: The celebration is an antidote to the extreme and sudden loneliness many of us envision as death. We want to gather all the people who loved us and knew us and maybe even admired us to accompany us one last time.

The people I encountered along the way did not grieve any less because they chose to celebrate a life even as they mourned a death. If anything, their grief seemed more intense to me, perhaps because it was not cordoned by stiff tradition and regimented ceremony.

This is not to say they eschewed ritual. It seems to me that even the most forward-thinking, tradition-despising survivor finds some value in ceremony. The ritual may involve religion, or it may not. It may involve deeply rooted historical rites, or it may not. It may involve professional facilitators, or it may not.

Here's how I see it. The new American way of death is personal, spiritual, and emotional. It is altruistic, futuristic, and individualistic. It can make the living appreciate the here and now.

As I conclude this tour, my mother is in treatment. She is fighting with the same ferocity with which she loves us, and we are hoarse with cheering. Searching for the fun in funerals, I found the if in life. The people I met along the way who had lost loved ones knew of the if. But in the passionate and disparate ways they memorialized their dead, I saw that Euripides was right: *Never that which is shall die.*

If—when—I lose someone I love, I hope I can remember that.

Last Will and Testament

I, Lisa Takeuchi Cullen, being of somewhat sound mind and body, would hereby like to bequeath the following:

Theresa Park, agent extraordinaire, who took me on as a charity case and became my adoptive big sister: my library of death-related reading.

Herb Schaffner, my wise editor, cheerleader, sounding board, friend: a *Godfather* poster. (To Marion Maneker, who took a chance on this book, and Joe Tessitore, who marshaled the troops behind it, my gratitude.)

Jan Simpson, editor at *Time*, who unwittingly kicked off this book by handing me a fun, if weird, assignment: my admiration. My bosses and colleagues, who pretended not to notice my unusually long maternity leave, and kept my name on my office door for its duration (though my computer was mysteriously "lost"): extra Saturday duty.

The end-trepreneurs, funeral directors, organizations, and schools I interviewed, who dealt patiently with my questions and presence: an end to my questions and presence. Lou Stellato and his family in Lyndhurst, New Jersey, who gave me particularly generous access to staff, services and clients: my warm regards. The National Funeral Directors Association, which allowed me to attend their events: my thanks.

The families and individuals profiled in the book who allowed me into their lives, shared with me their grief and laughter, and blessed me with stories and memories of their loved ones: I owe you the soul of this book.

To the following people, I bequeath my love: my sister Emy Seeley, who delivered hot meals upon deadlines and drove with me to Ocean City, New Jersey. My brother George Reilly, who babysat, unasked. My other brother, Ken Reilly, who put aside mortal fears to spend the night next to our grandfather's corpse. My oldest friend, Kristin Brueggeman Chaudry, who housed us in Denver; Brett Sterrett in Salt Lake City; my cousin Gabby Heintzelman in Tokyo; Mark Hudis and Natalya Preiser in Los Angeles; Richard and Gillian Reilly in San Francisco.

Lots of people provided my daughter with love and care: the saintly Melissa Hansen, who watched over Mika in Salt Lake City; my husband's aunt, Rita Valenti, in Westminster and Atlanta; my mother-in-law, Nancy Cullen, who kept Mika in hugs; my surrogate mother, Marlene Kahan, who kept her in toys; our babysitter, Carol Foster, who kept her jolly and safe.

For my mother, Hiroe Takeuchi Reilly, I wish nothing less than life. To her I owe the strength to see this through. I spent the close of 2005 not finishing this book as I was meant to, but by her side as she fought to survive her cancer. As I prepared to leave the hospital after a particularly wretched day, she called me to her bed.

"You may think you are alone," she said, holding my shaking hand, letting me cry. "You are not. You are never alone."

She was right: I have my husband. Each of these kindnesses from each of these people, I will not forget. But my husband, Christopher P. Cullen, provided all of the above on an hourly basis. He built me the attic lair where I write. He conquered the poopy diaper and the Elmo impersonation. He scheduled practice times during her naps and gigs around my deadlines. At my nadir, he carried me. To my husband I bequeath a laser-etched, black marble tombstone engraved with an ode to our love.

No, not really. Hon, you'll have to make do with a plasma TV.

Index

Page numbers in *italics* refer to illustrations.

The Wonders
of the Kingdom

A. Alan Alaimo

Unless otherwise indicated, Scripture quotations used in this book are from The King James Version of the Bible.

Other Bible versions used include:
New International Version (NIV) Copyright © 1973, 1978, 1984, International Bible Society. Used by permission of Zondervan Bible Publishers.

The Living Letters, (TLB) Copyright © 1971 by Tyndale House Publishers.

An effort has been made to locate sources and obtain permission where necessary for the quotations used in this book. In the event of any unintentional omission, modifications will be gladly incorporated in future editions.

Pathway Press

ISBN 0 87148 9880

The thoughts expressed in this book are not
necessarily those of Pathway Press.

Having known Reverend A. Alan Alaimo for many years as a co-worker and friend in the Gospel, he is qualified in his commitment in building God's kingdom through print. An honored leader of the Church of God as a successful Pastor, District Overseer, Chairman of the State World Missions Board and member of the State Council. His dedication, love and humility in his writing will glorify the Christ he serves.

Reverend Thomas Grassano
Pennsylvania State Overseer
Church of God, Cleveland, Tennessee

Alan Alaimo is a practical visionary. That is, he makes his visions come to pass. I first met him when I came to the USA from Great Britain in 1956. He was a true friend in time of need. From then until now I have seen his visions materialize - in evangelism, pastoral work, church building and other ways. Now he has written books. They deserve to be read because what Al decides to do, he does well. I comment this reading to as wide a readership as possible. It will bless you.

David B. Owen
Former Assistant Supt. of the Penn/Del District - Assemblies of God
Retired Associate Professor North Central Bible College, Minneapolis, MN

Alan Alaimo pastors the prestigious Full Gospel Church of God in Camp Hill, Pennsylvania. His vision and burden to reach those who are without Christ is not confined to his city, state or country. Pastor Alaimo embraces the world as his parish. Thus, he has led the Camp Hill Church to be concerned about people around the world who do not know Jesus Christ as Savior and Lord. Pastor Alaimo and his congregation faithfully support missionaries, national workers and projects in many areas of the world. They are truly a church with a heart for missions.

Lovell R. Cary
Assistant Director Church of God World Missions
A Missionary to the Philippines for 20 years

Al Alaimo has been a personal friend for over 30 years. He has always remained true to the fundamentals of the Gospel. His ministry is . . . motivated by a passion for the lost . . . moved by a compassionate heart . . . marked by dynamic preaching.

Pastor Paul Wislocky
Christian Life Assembly
Camp Hill, Pennsylvania

Reverend Alan Alaimo is a treasured friend and brother in Christ to The Couriers. Since the 1950s when he first invited us to sing and minister to his congregation in Bethlehem, Pennsylvania, we bonded to him. This dear man of God has survived and thrived in the very face of strong adversity. His diligence and tenacity are gifts of God to the Church. I cannot recall a conversation with him on any subject when he didn't somehow manage to speak of his vision for the church. It's his calling, his focus and his passion.

Neil Enloe
Song writer and Gospel singer
The Couriers, Harrisburg, Pennsylvania

Pastor Alan Alaimo and his wife Mary are wonderful servants of the Lord. Their ministry and warm fellowship have enriched my family and me. Alan Alaimo hears from the Lord and I believe you will be blessed by reading what he has written.

W.A. Davis
Director of the Western School of
Christian Ministry, Fresno, California

The Wonders of the Kingdom

God Is At Work Among His People

PRAISE
FAITH
HEALING
PRAYER
HOLY SPIRIT

Lovingly dedicated

To my godly parents
Charles and Anna Alaimo
who led me to Christ and taught me
by their life and words what it means to walk
the Christian path. I am eternally grateful
for their persistence with me in my developmental years,
and for revealing to me "The Wonders Of The Kingdom"
through example.

~~~~~~~

Affectionately dedicated

To my wife
Mary
my best friend
and special gift from God.

PREFACE

The desire to write about "The Wonders of the Kingdom" came to me when I realized that many Christians believe in the tenets of their faith but do not fully understand them. It is one thing to have a belief; it is something different to understand what you believe and be able to communicate that belief to others. We live in a world filled with wonders such as the miracle of creation, the birth of a baby, and the wonderful things that happen to us each day. I thought of Niagara Falls, the Grand Canyon, and many other places we like to visit. We are amazed at the wonders of this century, like airplanes, space ships, television and the computer.

I am fascinated with the most wonderful wonders of all; the wonders of the spiritual world and those are the wonders that are in this book. *The Wonders of the Kingdom* is divided into three volumes.

The first volume reveals that *God Has A Plan To Bless His People* and He has given us the wonders of Salvation, the Cross, the Resurrection, Christianity and Christ's Return.

The second volume reveals that *God Is At Work Among His People* giving us the privilege to participate in the wonders of Praise, Faith, Healing, Prayer, and the gift of the Holy Spirit.

The third volume reveals that *God Will Give A Choice To His People* through the wonders of Jesus Christ, Heaven, Hell, Love and Tithing.

My spiritual gift is the spoken word. I write to speak and do not speak to write. Therefore, the transition from pulpit to paper was a tedious and difficult one for me. Split infinitives, mixed metaphors, dangling participles, and very long sentences had to be eliminated. These volumes were originally prepared from material intended only for the pulpit. The material in these chapters has been compiled from sermons delivered to the Full Gospel congregation in Camp Hill, Pennsylvania. I am indebted to this family of believers for their enthusiastic support of the preaching of God's Word. For this reason, at times the identity of quoted material and the proper credit may not be fully given. A scrupulous effort has been made to give such acknowledgments wherever possible. Acknowledgment is generously and gratefully given to the hundreds of great preachers that I have heard and the hundreds of books that I have read whose authors have given me insight to the wonders of the kingdom.

I want to acknowledge the work of Carmen Faust Miller of Carmen's Computer Graphics, for her excellent computer layout experience that provided a camera-ready manuscript. Her work is deeply appreciated.

It is my prayer and desire that the messages will be understood and applied and you will *"grow in grace, and in the knowledge of our Lord and Saviour Jesus Christ. To him be glory both now and for ever, Amen"* (2 PETER 3:18).

A. Alan Alaimo

October 1, 1997

CONTENTS

PART ONE

THE WONDER OF PRAISE

Psalm 150:1-2

True worship and praise involve a wonder. We are living in an unusual period of human history, when the knowledge of humans has increased in technology, in science and in physical development. In this generation, the astronomers have explored the heavens and seen planets that are beyond the human eye. We have watched rockets and space shuttles being launched into outer space. We have shared the journey with our astronauts as they landed and walked on the moon. Thanks to TV documentaries, we have seen everything from the conception of a baby, to the eruption of a volcano. We have watched flowers grow and fish spawn. It is because of television and computers that the mysteries of life are no longer mysteries. The famous seven wonders of our world are no longer wonders. They are only tourist attractions. In the heart of human beings there is still a longing to seek and experience things greater than themselves that will produce a wonder.

What is a wonder? Many different words cluster around the idea of a wonder, such as the words "amazement," "surprise," "astonishment," "bewilderment," "admiration," "awe" and "fascination," just to name a few. We must recognize the fact that true wonder is not a passing emotion or some kind of shallow excitement. It has depth to it. A true wonder reaches right into your heart and mind and shakes you. It not only has depth but it also has value and interest. A wonder will enrich your life and bring you excitement. We need to return to the wonder of praise! Why do we praise Him? We praise Him because we are amazed at the greatness of our God.

Praise has always had a part in the worship of the Lord's people. The psalmist David touches the heart of the wonder of praise. *"I will make thy name to be remembered in all generations, therefore shall the people praise thee for ever and ever"* (PSALM 45:17).

You cannot remove the element of praise in worship. In the Hebrew, *praise* is used over 200 times, 45 times in a hymn of praise. Praise is also used 45 times in worship with extended hands, and 100 times in celebration. Over and over again we see the encouragement in God's Word to praise the Lord. Praise will turn your life around and will put new meaning to the things you do each day. It ought not to be Thanksgiving once a year for the Christian, but it ought to be *thanksliving* every day.

Praise turned the tide for Paul and Silas when they were in prison. The jailbirds became songbirds and God sent angels down to sing bass with an earthquake. No jail has been built that can contain the praise of God. We can have a song in our heart in the night and see our praise go up through the ceiling, to the sky, and up to the throne of God. The moment our praise touches the presence of God, something wonderful always happens. Praise is not "doing" something for the Lord, but it is the presenting of your soul and spirit to Him. In other words, it is the giving of your inner spirit so completely that you know you are one with the Father. Praise unites you with God. Praise brings a spiritual endowment within you. Praise opens up a relationship with God.

There is an old expression, "God will give it to you, if He can get it through you!" We need to understand what causes us to praise the Lord. There is a difference between the Christian who does not praise God and the one who does. I have observed that when praise is the lifestyle of a Christian, their eyes of faith are on the Lord. They see the hand of God in everything they do. What makes a Christian's lifestyle one of praise? Perhaps it is because some Christians have eyes of faith fixed on the Lord while others look only to themselves. When God is the center of your life you can praise Him every day because you will always find blessings no matter how difficult your circumstances.

The Wonders of the Kingdom

ONE

The First Wonder of Praise

GOD HAS GIVEN US HIS NAME

T he psalmist David expresses the scriptural position of the name that should be praised: *"That men may know that thou, whose name alone is JEHOVAH, art the most high over all the earth"* (PSALM 83:18). All the way through the book of Psalms, even in the passages that display grief, you feel, as someone expressed, like a smoldering volcano of praise ready to burst out at any moment into a great flame of gratitude to God. David's secret in developing a lifestyle of praise was that He knew the name of God to be a special name: *"I will extol thee, my God, O king, and I will bless thy name forever and ever. Every day will I bless thee and I will praise thy name forever and ever"* (PSALM 145:1-2).

When Jesus made His triumphant entry into Jerusalem, there was a spontaneous reaction among the people. It was not orchestrated by the disciples of Jesus, it was something that came from the innermost depth of the human heart. *"And the multitudes that went before, and that followed, cried, saying, Hosanna to the son of David: Blessed is he that cometh in the name of the Lord; Hosanna in the highest"* (MATTHEW 21:9). The people entered into a spirit of praise because they recognized that Jesus came in the name of the Lord. We praise God because of *His name*!

One of the things that we need to observe about the Old Testament event of the burning bush in Exodus chapter 3 is that the burning bush represents God. In addition to the fact that there was a strange physical phenomenon of a bush burning that was not consumed, it was even greater that God gave us a Name to Praise. The name of God should be praised every day of our lives. It was in the burning bush experience that Moses witnessed the presence of God: *"And Moses said unto God, Who am I, that I should go unto Pharaoh, and that I should bring forth the children of Israel out of Egypt? And he said, Certainly I will be with thee; and this shall be a token unto thee, that I have sent thee: When thou hast brought forth the people out of Egypt, ye shall serve God upon this mountain. And Moses said unto God, Behold, when I come unto the children of Israel, and shall say unto them, The God of your fathers hath sent me unto you; and they shall say to me, What is his name? what shall I say unto them? And God said unto Moses, I AM THAT I AM: and he said, Thus shalt thou say unto the children of Israel, I AM hath sent me unto you. And God said moreover unto Moses, Thus shalt thou say unto the children of Israel, The LORD God of your fathers, the God of Abraham, the God of Isaac, and the God of Jacob, hath sent me unto you:* **this is my name for ever**, *and this is my memorial unto all generations"* (EXODUS 3:11-15). God gave us a name that is above all names. Just the mention of His name should cause us to praise Him spontaneously with excitement and with zeal. God represents His essential nature by the name He calls Himself.

God is called the "All-powerful," *Elohim*. The name *Elohim* contains the idea of creative and governing power, of omnipotence and sovereignty. This is clearly indicated by the phrase in Genesis 1:1, *"In the beginning God!"* The Hebrew word from which this word "God" is translated is *Elohim.* It is *Elohim,* who by His mighty power created the vast universe, who said, *"Let there be..."* And by His spoken word *"the worlds were framed by the word of God, so that things which are seen were not made of things which do appear"* (HEBREWS 11:3). It is most appropriate that by this name God should reveal Himself as Elohim and bring cosmos out of chaos, light out of darkness, habitation out of desolation, and life in His image.

God is called *Adonai,* "my Lord." The name *Adonai* translated, "Lord" signifies ownership or mastership. God gives a classic illustration about the obedience that should be in His name *Adonai: "A son honoureth his father, and a servant his master: if then I be a father, where is mine honor? and if I be a master, where is my fear? saith the LORD of hosts unto you, O priests, that despise my name. And ye say, Wherein have we despised thy name?"* (MALACHI 1:6). There are hundreds of references in the New Testament that teach us to praise God because of His Lordship over His children. We belong to God who is our Lord and Master. We are instructed to glorify God in body and spirit, which are His (I CORINTHIANS 6:19-20).

God is called "The Almighty," *El-Shaddai.* The word "Almighty" does suggest the all-powerful, the mighty, the power to be able to do anything and everything at any time. God's name not only demands respect, but is the means of our life of praise. The name *El-Shaddai,* or "Almighty God" speaks to us of the inexhaustible provisions of His bounty, of the riches and fullness of His grace. We praise His name *El-Shaddai* because it means that from God comes every good and perfect gift, and that He never wearies of pouring His mercies and blessings upon His people.

The name that is used most supremely is the name God

gave to Moses in the desert. *"This is my name forever." "I AM THAT I AM" "He is Jehovah - The I Am."* The meaning carries the following idea: *I Am that which I will be.* God is saying, I Am for you precisely what you need me to be. **I Am** means the Eternal One. **I Am** means He is sovereign. **I Am** means He is faithful. Its origin is exactly the same as that of *Jehovah* which denotes the One who will always be personal, continuous, and absolute in His existence. In the King James Version of our Bible the Hebrew word *Yahweh* is translated JEHOVAH or LORD in capitals to distinguish it from another Hebrew word. *"And God spake unto Moses, and said unto him, I am the LORD: And I appeared unto Abraham, unto Isaac, and unto Jacob, by the name of God Almighty, but by my name JEHOVAH was I not known to them"* (EXODUS 6:2-3).

We find in His name a reason to praise God. We are instructed to *"Enter into his gates with thanksgiving, and into his courts with praise: be thankful unto him, and bless his name"* (PSALM 100:4). There are eight names compounded with the name "Jehovah" in the Old Testament. When God wished to give a special revelation of Himself, He used the name Jehovah (Hebrew Yahweh,) translated "Jehovah."

He is Jehovah-tsidkenu - *The Lord Our Righteousness*
(Jeremiah 23:5-6).

He is Jehovah-m'kaddesh - *The Lord Who Sanctifies*
(Leviticus 20:7-8).

He is Jehovah-shalom - *The Lord Is Peace*
(Judges 6:24).

He is Jehovah-shammah - *The Lord Is There*
(Ezekiel 48:35).

He is Jehovah-rophe - *The Lord Who Heals*
(Exodus 15:22-26).

He is Jehovah-jireh - *The Lord's Provision Shall Be Seen*
(Genesis 22:1-14).

He is Jehovah-nissi - *The Lord My Banner*
(Exodus 17:15).

He is Jehovah-rohi - The Lord My Shepherd
(Psalm 23:1).

God is saying to us today: Praise Me! I Am Jehovah God, the great I Am. I Am your Provider, I Am your Righteousness, I Am your Shepherd, I Am your Peace, I Am your Healer, I Am your Banner, I Am your Sanctifier, I Am your Promise and I pledge to be there always for you.

There is a name that God has given us whereby we can praise Him. *"Thou shalt call his name JESUS: for he shall save his people from their sins."* His name means: *"Emmanuel, which being interpreted is, God with us"* (MATTHEW 1:21,23). Why do we praise Him? At the name of Jesus, our Emmanuel, demons tremble! At the name of Jesus I am not lost anymore! At the name of Jesus I live a life of freedom from sin! God has given us the understanding of His name so that we can intelligently praise Him. The wonder of praise is of the name of God. *Praise ye the Lord*!

TWO

The Second Wonder of Praise
GOD HAS GIVEN US HIS WORD

The triumphant entry of Jesus coming into Jerusalem riding on a colt and receiving Praise was prophetically recorded in Zechariah 9:9. Getting the colt was a unique experience. Jesus said: *"Go ye into the village over against you; in the which at your entering ye shall find a colt tied, whereon yet never man sat: loose him, and bring him hither. And if any man ask you, Why do ye loose him? thus shall ye say unto him, Because the Lord hath need of him"* (LUKE 19:30-31). If anyone makes an inquiry of your actions, tell them that you have a **word from the Lord**. The major thing that causes us to respond to praise and worship to the Lord is the Word that is within our hearts not just any word but the words from God's Holy Word. Do you know that if you would read the Bible, and adhere to it, and respond to it, and claim it, and live in it, you would find yourself praising the Lord. There is something transforming about the Word of God. The life of God comes within us through the word!

The Bible is an open book. When you are lonely, there is a promise. When you are sick, there is a word of deliverance.

When you are down, you can be lifted up. True praise is given to God when you open His Word and find that your need will be met. In God's Book you will discover His gifts. We must open the Book to find spiritual revelations, revealing truths that set us free, and words of encouragement. It is only when the Word of God comes alive that there will be a natural response of praise to God.

The Word was described by the psalmist David when he said: *"Thy word is a lamp unto my feet, and a light unto my path"* (PSALM 119:105). When we praise we are responding to the power of the Word. It is not a mere emotional experience. The power of the Word of God burns in our heart. It was the expression of the prophet Jeremiah when he said: *"His word was in mine heart as a burning fire shut up in my bones"* (JEREMIAH 20:9).

There is a religious movement today that is saying, "We do not need to read the Bible any more like we used to; all we have to do is to rely on the Spirit." They sit and wait for messages from the Spirit, and this has become a substitute for the Word of God. They claim that they do not need the preached Word, or the living Word, but only a current revelation of prophecy or a message by the Spirit. How sad it is when spiritual mentality deteriorates to the point that something good replaces that which is perfect. When the Bible is a closed book you will never respond to the wonders of praise. There is no substitute for the Word of God. The prophet Amos gives us a revelation of what happens when the Word of God is not important anymore. *"Behold, the days come, saith the Lord GOD, that I will send a famine in the land, not a famine of bread, nor a thirst for water, but of hearing the words of the LORD; And they shall wander from sea to sea, and from the north even to the east, they shall run to and fro to seek the word of the LORD, and shall not find it"* (AMOS 8:11-12).

Perhaps most of us know the story of the two bachelor

brothers named Robertson. One was a Doctor of Divinity and the other a Doctor of Medicine. Both the preacher and medical doctor lived together in the same house where they practiced their profession and were frequently mistaken for each other. One day a person who needed medical attention telephoned their home and not knowing which Robertson was answering the call said: "Are you the doctor who preaches, or the doctor who practices? I want the one who practices, please." We should be the Christian who preaches and practices, one who lives the Word of God, so that our lives will be a living praise to God. Paul said: *"Ye are our epistle written in our hearts, known and read of all men"* (2 CORINTHIANS 3:2). We are practicing what we are preaching. The words of God are words of action. It is not so much our preaching, but our living. We must allow the Word of God to have a proper place in our heart. The signs that follow a believer produce praise to the Lord our God.

The reason many people are not praising God today is because the Word of God does not live within their hearts. God knows if you will read His Word you will praise Him. There will come a glow, a witness an excitement in the activities of your life when you begin to take the Word of God and apply it to your life. The Word of God has life, power, victory and glory. Here is the wonder of praise. *"But ye are a chosen generation, a royal priesthood, an holy nation, a peculiar people; that ye should shew forth the praises of him who hath called you out of darkness into his marvelous light"* (I PETER 2:9). Praise God for His Word, it has power to do wonders!

The Wonders of the Kingdom

THREE

The Third Wonder of Praise
GOD HAS GIVEN US HIS PRESENCE

G od's presence causes me to praise Him, and when I praise God I feel His presence. *"And a very great multitude spread their garments in the way; others cut down branches from the trees, and strowed them in the way. And the multitudes that went before, and that followed, cried, saying, Hosanna to the son of David: Blessed is he that cometh in the name of the Lord; Hosanna in the highest"* (MATTHEW 21:8-9). The presence of Jesus was vital in creating an atmosphere of praise. It was the presence of Jesus that caused them to respond with shouts of "Hosanna in the highest." It wasn't just a day of celebration for those who followed, nor did they follow just because Jesus was going into Jerusalem to cleanse the Temple. It wasn't the march, nor the spreading of their garments in the way, it was Jesus Himself.

It is a wonder to me that God would condescend to come down among us so that we would have a reason to praise Him. *"And the Word was made flesh, and dwelt among us, (and we beheld his glory, the glory as of the only begotten of the Father,) full of grace and truth"* (JOHN 1:14).

Just imagine, we can feel the presence of God! What an honor! Praise is the gateway into God's presence. Isaiah 60:18 speaks of the gates of praise. *"Violence shall no more be heard in thy land, wasting nor destruction within thy borders; but thou shalt call thy walls Salvation, and thy gates Praise."* I am thankful to Christ Jesus for providing the solid walls of salvation that He has made possible through His death and resurrection. My life would not be complete without the *"gates of praise"* that swing open and allow me to enter His presence. It is time to oil the hinges of the gates of praise. For many people the gates of praise have been closed, and as a result they have been denied the presence of God. Isaiah went on to say that when the gates of praise are open, *"The sun shall be no more thy light by day; neither for brightness shall the moon give light unto thee: but the LORD shall be unto thee an everlasting light, and thy God thy glory"* (ISAIAH 60:19). All you have to do is to open the gate and He comes in!

In Dave Billington's worship and praise chorus, "Awesome In This Place" he expresses the need of first passing the gates of praise before we enter into God's presence. When we enter into God's sanctuary we will meet him face to face. When we look upon his countenance we see the fullness of his grace. God's presence makes every place awesome. He is worthy of our praise.

We need to learn how to recognize the presence of God in all things so that we can be instruments of praise for our Lord. Many people say, "I don't feel like praising the Lord." I have discovered a wonderful secret of praise. Though my physical body may feel exhausted, yet when I feel the presence of God I am renewed and I respond with praise. In order to feel the presence of God we need to be aware of God's goodness in all things. This is what the psalmist David meant when he said: *"My mouth shall speak the praise of the LORD: and let all flesh bless his holy name for ever and ever"* (PSALM 145:21). When you realize that God is accessible to us, that He walks and talks with us and tells us that we are His own, then you will praise Him. The moment this truth is revealed, you will sing the doxology written by Thomas Ken, "Praise God from whom all blessings flow; Praise Him all creatures here below; Praise Him above ye heavenly host; Praise Father Son and Holy Ghost. Amen" (Genevan Psalter).

Ezekiel the prophet was lifted out of himself, when he saw the likeness of the coming Messiah and the appearance of the fire around the throne of God. *"This was the appearance of the likeness of the glory of the LORD. And when I saw it, I fell upon my face, and I heard a voice of one that spake"* (EZEKIEL 1:28). It was after he saw the presence of God that he fell down before God in worship and praise. I have discovered a little secret in the lifestyle of one who praises the Lord. There is always an awareness of the presence of God in their lives. Recognize that God is alive, recognize that He is able to dwell among us, and when you do there will be a natural response of praise to our God. We need to be reminded that our Lord said: *"I will never leave thee, nor forsake thee"* (HEBREWS 13:5). He has given us a promise that His presence will always be with us. Remember and therefore be guaranteed of a life of praise.

At the dedication of the Temple during the reign of King Solomon the congregation witnessed the wonder of praise. *"When the priests were come out of the holy place, that the cloud filled the house of the LORD, so that the priests could not stand to minister because of the cloud: for the glory of the LORD had filled the house of the LORD. And he said, Blessed be the LORD God of Israel"* (1 KINGS 8:10-15). The glory cloud of God's *Shekinah* (His Presence) came down from heaven and filled the house. The wonder of praise is that it brings heaven's best down to earth, into our hearts and into our lives.

27

The Wonders of the Kingdom

FOUR

The Fourth Wonder of Praise
GOD HAS GIVEN US PURPOSE

One of the great lessons Jesus taught was that if people would refrain from the expression of praise, God would cause the stones to praise Him. This is how the triumphant entry of Jesus into Jerusalem is recorded in Luke 19:36-41: *"And as he went, they spread their clothes in the way. And when he was come nigh, even now at the descent of the mount of Olives, the whole multitude of the disciples began to rejoice and praise God with a loud voice for all the mighty works that they had seen; Saying, Blessed be the King that cometh in the name of the Lord: peace in heaven, and glory in the highest. And some of the Pharisees from among the multitude said unto him, Master, rebuke thy disciples. And he answered and said unto them, I tell you that, if these should hold their peace, the stones would immediately cry out. And when he was come near, he beheld the city, and wept over it."* What made Jesus weep? He saw the great religious city of Jerusalem preoccupied with a form of worship but denying the praise that belonged to the Son of God.

God has given us a purpose to praise. He created us in His image and likeness, (GENESIS 1:27). We were created as a child of

God, to respond to God, honor God, love God, live for God and serve God. This is our purpose and for this reason we praise Him.

We need to see praise as an opportunity to express our feelings to God. David said: *"Oh that men would praise the Lord for his goodness and for His wonderful works to the children of men!"* (PSALM 107:8). It should be a natural instinct to praise the Lord. The Bible says *"Let every thing that hath breath praise the LORD. Praise ye the LORD"* (PSALM 150:6). Our main purpose to be on planet Earth is to worship and serve the Lord. It is an eternal purpose given to us by God our creator so that our lives would be complete on earth. For years as an unbeliever we praised and glorified the deeds of the flesh, but we realized, as Solomon did, that it was all vanity. Now our eyes are open to the purpose of God. Let us *"Enter into His gates with thanksgiving, and into His courts with praise; be thankful unto Him"* (PSALM 100:4).

Praise brings victory! There is a marvelous account of the purpose of praise in the life of King Jehoshaphat recorded in 2 Chronicles 20:15-22. The armies of several nations were coming against the children of Israel to destroy them. King Jehoshaphat was assured by God: *"Be not afraid nor dismayed by reason of this great multitude; for the battle is not yours, but God's."* He was instructed by God to appoint singers and they sang: "Praise the LORD; for his mercy endureth for ever." We see clearly the purpose of praise. *"And when they began to sing and to praise, the LORD set ambushments against the children of Ammon, Moab, and mount Seir, which were come against Judah; and they were smitten."* When praise is present, the Lord will send an ambush before the enemy. The purpose of praise is to bring down the stronghold of the enemy. When we experience in our lives that God is working to protect us, how can we remain silent?

God gave all of the ten lepers in Luke 17:12-19 a purpose

to praise Him; but only one returned to give thanks. Jesus healed all of them and told them all to show themselves before the priest. They all had a purpose to go back to Jesus to thank Him, but only one saw the wonder of praise. Possibly in the lives of the nine lepers who were healed they had purposes that they felt were greater than the purpose of going back to thank Jesus. Maybe they had a desire to be united with their families again, or possibly a desire to go back into the marketplace that had been off limits to them. Here is a lesson to be learned. Let us not be distracted or preoccupied with purposes that are less important than God's desire for us. *"Oh that men would praise the Lord for His goodness and for His wonderful works to the children of men"* (PSALM 107:8) . We were redeemed to praise the Lord.

5
FIVE

The Fifth Wonder
of Praise
GOD HAS GIVEN
US HEALING POWER

The healing effect of praise in our lives is a wonder. Praise soothes the nerves, clears the conscience, and provides peace for troubled times. Praise is the greatest tranquilizer in the world. Praise brings complete confidence that God is in control. It is in the created atmosphere of praise that we begin to see God high and mighty, the One who has the power over all the elements of this world. We should praise the Lord for healing before the healing process begins. It is easy to praise God after you have been healed but it take faith to praise Him before you are healed. As we begin to praise Him for who He is, the Holy Spirit will reveal the power that is available to us now through Jesus Christ. Praise is healing! It is the greatest therapeutic power in all the world. Here is our promise: *"But unto you that fear my name shall the Sun of righteousness arise with healing in his wings"* (MALACHI 4:2).

I received a phone call on Saturday night, April 13, 1996 at 10:45 that our daughter Brenda and son-in-law Walter and our three grandchildren were involved in a serious automobile accident. Cristy, our nine year old grandchild, was seriously

33

hurt and was taken to John Hopkins Hospital in Baltimore, Maryland. We arrived at the hospital three hours later to witness a great nightmare of life. It was difficult to see a family member physically damaged by a careless driver. We were told that the damage that was done to Cristy's left eye was a fracture of the inner bone which supports the eyeball . During the middle of the night as we sat with a discouraged heart in her room waiting for the doctors to operate on Cristy's eye, I discovered again the wonder of praise. I began to look for reasons to praise the Lord. I wanted to see God in this situation. I began to review the accident with Walter and Brenda: When they took Cristy out of the car, she was unconscious. While they were waiting for the ambulance to come, in the providence of God a helicopter from the trauma center in Baltimore was on its way to gas up in Frederick, Maryland. The helicopter was in the sky above them during the time of the accident. The police requested assistance, permission was granted, and in six minutes Cristy was in John Hopkins Hospital. Again, in the providence of God, a most honored eye specialist performed the delicate operation on Cristy's eye. She was cared for by a team of doctors from the Wilmar Eye Institute, located in John Hopkins Hospital. As we looked for things to praise the Lord for in this accident we were overwhelmed that God was there all the time. We began to praise God for the helicopter being at the right place at the right time. We thanked God for care from the best doctors in the world, and for several other wonderful events of that evening. As praise dominated our troubled hearts, we were all lifted up with confidence that God was going to bring full restoration. We watched our nine year old granddaughter become strong in her faith that she would be better again. Look for things to praise God for during your troubled times and praise will carry you over the tragedy, bringing healing and restoration.

Praise is healing—physical, spiritual and emotional. The Bible encourages us, *"Praise ye the LORD: for it is good to sing praises unto our God; for it is pleasant; and praise is comely. He healeth the broken in heart, and bindeth up their wounds"* (PSALM 147:1,3). There is power in praise. Do not underestimate the power of praise! We need to live in an atmosphere of praise to God for our protection and the protection of our children. Many times the healing process from God is being done without our knowledge, and for this reason praise should dominate our lives. Thank Him for the healing power that is available to you.

I have discovered something about human nature that God desires to change. It may be true that there are four or five things that are physically wrong with you, but there are more than twenty things that are right with you. As you begin to praise the Lord for the things that are right, the four things that are physically wrong will become three. You will then discover that because of praise and thanksgiving the three things which are physically wrong will become two. This is how praise will demonstrate itself with the power of healing and deliverance.

A classic example of praise in the life of a person is Hannah. She desired a child but was barren. She was hurting, and needed healing. God answered her prayer. Listen to the expression of praise as she leaves her miracle child Samuel in the Temple with Eli the High Priest, *"My heart rejoiceth in the Lord, mine horn is exalted in the Lord, my mouth is enlarged over mine enemies, because I rejoice [praise] in thy salvation"* (1 SAMUEL 2:1). It wasn't easy for Hannah to leave her only child with Eli the High Priest. With a hurting heart she was fulfilling her vow to God. Hannah expressed praise to the Lord. She had a rejoicing heart. Her strength was lifted up, and her mouth was enlarged over her enemies. She was determined to praise the Lord.

35

Our enemy needs to know that we are not fearful people. It is through praise and worship that we can rise above our obstacles and weaknesses. We must dwell on good things and sing the song of rejoicing. When we gather in corporate praise with a group of people that are hurting, sick and in need, it is the wonders of praise that will bring deliverance. The wonder of praise is a healing power.

6

The Sixth Wonder of Praise

GOD HAS GIVEN US THE WILL OR CHOICE TO PRAISE

It is a wonder how even in the middle of trouble and death, there can be the expression of praise to our Lord. Praise is an action of the will. When you will to praise God you will! God does not force you to praise Him. A problem in our praise life is the vacillation of our emotions. There will be times when you will not feel like praising the Lord. I am amazed that Almighty God who has given us so many benefits allows it to be our choice to praise. Each person has the capacity to know, to feel and to decide whether or not to praise our creator. If we let emotions determine our praise life we will certainly fail. Thank God, emotions do not determine our praise life. We have made a choice that is based on our relationship with the Lordship of Jesus to give God thanks for all things.

Praise must be an act of our will. When you choose to praise in spite of what you are going through, you must condition your mind, heart and body. In the natural you do not feel to respond to God, but when you set your will to praise God your body will submit. Praise is subject to our free will. Paul states this reality: *"Be joyful always, pray continually; give*

37

thanks in all circumstances, for this is God's will for you in Christ Jesus" (1 THESSALONIANS 5:16-18 NIV). If it is the will of God, then it should be our will. Jesus alluded to this truth when he said: *"Thy will be done in earth, as it is in heaven"* (MATTHEW 6:10). If you know that it is the will of God to praise, then there should be an instant response.

God does not tell us to always feel grateful, but He does, however command us to always give thanks. Even when we do not feel emotionally grateful, we can still choose to give thanks and praise to God. His desire is, just love me, just praise me, just honor me, whether or not you feel like doing it. The psalmist David said: *"My heart is fixed, O God, my heart is fixed: I will sing and give praise"* (PSALM 57:7). The just shall not live by feeling, the just shall live by faith. We have been born to serve the Lord.

SEVEN

The Seventh Wonder of Praise

GOD HAS GIVEN US TRIUMPH OVER THE DEVIL

It is a wonder how praise makes the devil tremble. It is a wonder how praise confuses the attacks of Satan and causes him to be defeated. Paul encouraged the Corinthians to offer up praise for spiritual victories in their lives. *"Now thanks be unto God, which always causeth us to triumph in Christ"* (2 CORINTHIANS 2:14) . We need to understand that the key and secret that causes us to triumph over the enemy of our soul and body is to praise God. There are people who are constitutionally negative, always more conscious of the dark clouds than of the blue skies. To these people the fountain of praise and gratitude is dried up and Satan will take advantage of this opportunity and keep that person perpetually defeated. We can reverse this trend by creating a lifestyle of praise.

Let us offer thanks to God for victory over the evil that is around us. The Romans learned from the apostle Paul in the seventh chapter of Romans that in his personal life there was an inner struggle. He said: *"For the good that I would I do not: but the evil which I would not, that I do"* (ROMANS 7:19). Paul was honest. He tells about human weakness. He was revealing that we are all involved in spiritual warfare. He goes on to say: *"I find then a law, that, when I would do good, evil is present with me. For I delight in the law of God after the inward man: But I see another law in my members, warring against the law of my mind, and bringing me into captivity to the law of sin which is in my members"* (ROMANS 7:21-24). Paul brings this chapter to a close with a question and then an answer. The question is: *"O wretched man that I am! who shall deliver me from the body of this death?"* (ROMANS 7:24). He has the answer for himself and for us when he says: *"Thanks be to God through Jesus Christ our Lord"* (ROMANS 7:25 NIV) . We praise God for the victory through Jesus Christ our Lord. Paul begins the eighth chapter with the declaration for victory and praise. *There is therefore now no condemnation to them which are in Christ Jesus, who walk not after the flesh, but after the Spirit. For the law of the Spirit of life in Christ Jesus hath made me free from the law of sin and death"* (ROMANS 8:1-2) .

It is a wonder that someday every knee will bow and every tongue will confess that Jesus is Lord. Even Satan will bow to the Lordship of Jesus. James gives us practical advice on how today we can achieve victory over the devil through the effectiveness of praise. It is the philosophy for life. *"Submit yourselves therefore to God"* (JAMES 4:7) . It means we should praise Him first and then: *"Resist the devil, and he will flee from you"* (JAMES 4:7) . Many Christians are trying to resist the devil without first submitting themselves to God in praise. We have it in reverse and that is why we have our spiritual problems. We are thinking if we get Satan out we will bring God in. The truth is that when we bring God into our lives through praise and worship, Satan must exit! When your heart is filled with praise to the Lord, there is no room for anything else. Satan must go! *"What shall we then say to these things? If God be for us, who can be against us?"* (ROMANS 8:31). There is a wonder, a power and a secret in praise when it becomes a way of life for the Christian.

There was a believer in God who was visited by trouble. He had one terrible blow after another coming down on him until his whole life was staggering. His family was taken, his wealth was gone, his health was shattered and his world had crumbled. I want you to see Job with a set face, a clenched hand, and a bent head, speaking these powerful words. *"Naked came I out of my mother's womb, and naked shall I return thither: the LORD gave, and the LORD hath taken away; blessed be the name of the LORD"* (JOB 1:21). And then there was a dead silence. For a moment everything was in the balance. I believe in that silent moment it was as if the unseen world and the angels were waiting for what would come next. Everyone was wondering and waiting as he was fighting his battle against the enemy of his soul. Then suddenly breaking the silence, with half a sob and half a shout, came the declaration of praise. *"Blessed be the name of the Lord, forever"* (JOB 1:21). Job was experiencing disappointment and pain, yet he expressed a positive statement. *"May the name of the LORD be praised"* (JOB 1:21 NIV) . If you want to know how Job turned darkness into light and a body of misery into a beautiful blossoming person—then observe his reaction to his trial. Job reached the conclusion that the Lord gives and the Lord takes away. These things were out of his control, but what was in his control was the privilege to praise and bless the name of the Lord. His faith found the victory over the devil.

The victory of the believer is to be able to praise God in everything. As we look to the example of Jesus during his days of trials before the crucifixion, He took the bread and the cup which were symbols of his life and death, and He then gave God thanks. Jesus went out from the upper room to Gethsemane to fight His great battle with Satan, and the Bible says He first led His disciples into a declaration of praise. *"And when they had sung an hymn, they went out into the mount of*

Olives" (MATTHEW 26:30). Here is the wonder of praise. In the most difficult hour in the life of Jesus, when He knew that one of His disciples was going to betray Him and He would be crucified on a cross, He entered into praise and worship. My blessed Lord sang before His battle with Satan. Jesus not only won the battle, He won the war. He triumphed for us.

PART TWO

THE WONDER
OF FAITH

Matthew 21:18-22

O ne of the chief difficulties in our Christian life is our
failure to see that faith can be received only as it is
imparted to the heart by God Himself. The apostle Paul
tells us *"God hath dealt to every man the measure of faith"* (ROMANS
12:3B). The Bible also says, *"So then faith cometh by hearing, and
hearing by the word of God"* (ROMANS 10:17). You cannot buy it with
money. You must believe a promise and at the same time you
must know how to appropriate the promise by faith. We have
formed many habits in trying to appropriate the promise by
belief, but belief is a spiritual quality. We do not try to believe
ourselves into an experience. Faith does not come through a
mental assent of God, it is something you experience in your
heart. Faith is spiritual. Faith is warm and vibrant, it lives and
throbs within the heart of a believer. The power of faith is
irresistible when it gets into your heart because it imparts unto
you an experience. This experience can only come from God. It
becomes a gift to us. *"Without faith it is impossible to please God"*
(HEBREWS 11:6).

God has asked you to receive and treasure His Words in
your heart. This becomes the act of faith. With God, speaking

and doing always go together; the deed follows the word. *"Shall He say it, and not do it"* (SEE NUMBERS 23:19). Scriptures declares, *"the LORD will do the thing that he hath spoken"* (2 KINGS 20:9). When I have a Word from God in which He promises to do something, I can always be certain that He will do it. I must simply hold fast to the Word and wait upon God. God will take care of fulfilling His Word. He has the power to do it, all we need is the faith to believe it. The operation of faith in our hearts, is unique. Before I ever feel or experience anything, I must hold fast to the promise, and I know by faith that God will make it happen.

What is faith? It is none other than the certainty that what God says is true. When God says that something exists or is, then faith rejoices, although it sees nothing of it. When God says that He has given me something, I know by faith with certainty that it is mine. When God says something will happen, then it will happen. When God says that something shall come to pass, or that He will do something for me, faith makes it real, just as real as if I had seen it. Faith will ask only for what God has said, and then rely on His faithfulness and power to fulfill His Word.

This is precisely what Paul stated when he was in a ship during a destructive storm and when everyone was fearful for their lives. *"Be of good cheer, for I believe God, that it shall be even as it was told me"* (ACTS 27:25). This is the secret of faith. We believe what God says.

Though it is difficult to explain faith, possibly this illustration may bring personal enlightenment. Imagine a prisoner being offered a pardon. He is overwhelmed and even in a daze after reading the document of his pardon. You may ask him. "Have you been pardoned?" His reply is "Yes." "Do you feel pardoned?" His answer "No." "But if you don't feel pardoned how can you know you are pardoned? You're not released from prison?" The prisoner points to the document. The document is the pardon that was given to him by the Governor. "This tells me so." What is faith? Faith is ability to point to the document. Faith is to believe and accept the truth; for it is the truth that sets us free. It is like this prisoner who felt nothing and had experienced no release, but he knew that the document was true! There is something inside of us that says God is going to work in mysterious ways, His wonders to perform, for His Word declares: *"Thou art the God that doest wonders: thou hast declared thy strength among the people"*(PSALM 77:14).

Let me share one more definition of faith. Faith is complete relaxation in the fact that what you are looking forward to has already happened.

The First Wonder
of Faith
THE CHARACTERISTIC
OF FAITH

"*Now faith is the substance of things hoped for, the evidence of things not seen*" (HEBREWS 11:1). Although Scripture does not define faith, a number of things are stated about it. "*Faith is a substance.*" To the natural mind, faith cannot be a substance. It is not tangible, you cannot see it or feel it with your physical body. In the apostolic times the word "*substance*" was used in the sense of a "*title deed.*" If you are the owner of a car and someone attempts to make a claim on that car, the question is settled by the one who has the title deed to the car. Faith is the firm assurance that God will do what He has promised. It is like money in the bank. God will do what He promised. We have the substance, the title deed. Through the eyes of faith we can see things that physically we may be limited to see. To the faith believer, what he sees has substance, it is not imaginary or a spiritual illusion. There is a faith in God, but there is also, the faith of God. Faith in God will take you to heaven, but the faith of God will bring heaven down to your soul. You need faith in God to be saved, but when you have the faith of God, you can move mountains and witness miracles.

Many Christians grow disillusioned in their Christian experience because God does not conform to their will. Faith does not insist that God conform to our ideas. Through the eyes of faith you see the big picture. You are God's plan and then conform to it because you know it is the best way. Faith says, "God, I believe the promise that you have given me." *Faith is the evidence* [literally, the proof] *of things not seen.* There is a witness of God's Spirit in our heart, that if God Almighty said it, I believe it. Hebrews 11:1 can also be described as *Faith is the title-deed of things hoped for, the sure expectation of things not seen.* When we seek healing for our body, by faith we hold the title deed, and claim the promises of God. Faith turns the promises of God into a substance so that we feel it, know it and have a witness of it.

Christian faith is trust and reliance on Jesus. Yes, this faith includes an assent to a statement of belief. We have all experienced this at conversion. But DYNAMIC FAITH is commitment, confidence in, and reliance on a person, Jesus Christ. I believe it though I don't fully understand it, because I know God is behind it. Here lies the power of faith, for it is not based on the characteristic of a church, a pastor, or an apostle. Faith is the life and nature of God within us. Many Christians are only creatures of emotion and not creatures of faith as God designed. We find ourselves as children of the flesh and not children of God. *"The just shall live by faith"* (ROMANS 1:17). We all have faith and confidence that the supernatural God will do wonders because He honors the Words of Jesus Christ. This is a wonder to me that through faith we can do even greater works than Jesus Christ because He is at the right hand of the throne of God making intercession for us.

After I spoke on the subject of faith in a neighboring church, a gentlemen came to me at the close of the service and gave me a definition of faith that has blessed me: "Faith is reaching up and taking hold of nothing and holding on until God turns it into something."

The Second Wonder of Faith

THE SOURCE OF FAITH

The source of faith is not the human mind. Many people want us to believe that faith is a mental experience. The source of true faith is God. We were taught that within the human body there are five sense organs that convey specific impulses to the central nervous system, they are sight, hearing, smell, taste and touch. When you were born again and became a child of God, you received what I call is the sixth sense. It is the life of God within you, it is the Spirit of God that comes alive within you; it is the presence of God that you feel, it is the relationship you have with God. Your spirit that was dead because of sin, is active as a result of accepting Jesus as the Lord of your life. *"There is therefore now no condemnation to them which are in Christ Jesus, who walk not after the flesh, but after the Spirit. For the law of the Spirit of life in Christ Jesus hath made me free from the law of sin and death"* (ROMANS 8:1-2). Faith is His life born and instilled within us that sees things as God sees them.

God created humanity in His image. This creation made men and woman partner with Him in the management of life. God gave man the ability to have authority with Him. He also gave man the potential of the mind, the ability to choose, and to see things as God sees them. There are people that teach faith through the power of the mind. I would like to share its weakness. Your mind is not the source of your faith. God is the originator of faith. Faith is not just a nice thing to have or something to make you unusually different. The truth about faith is that you cannot receive good things from God without faith. *"Without faith it is impossible to please God"* (HEBREWS 11:6).

There are those that teach possibility thinking, and that is powerful: Whatever you think and imagine you can have it. Possibility thinking sounds great, it has done a lot of good and helped many people. Possibility thinking is not faith, its source is not God, its source is the human mind. Let me illustrate where possibility thinking would not work. Sarah was 90 years old. All the organs in her body that could produce a life were dead. Abraham was 100 years old. He needed something beyond possibility thinking, he needed "impossibility thinking." Possibility thinking comes from man, impossibility thinking comes from God. This is where God comes in. This is how faith operates. The Bible expresses Abraham's faith this way: *"Who against hoped believed in hope, that he might become the father of many nations; according to that which was spoken, So shall thy seed be. And being not weak in faith, he considered not his own body now dead, when he was about an hundred years old, neither yet the deadness of Sarah's womb. He staggered not at the promise of God through unbelief, but was strong in faith, giving glory to God. And being fully persuaded that, what He had promised, he was able also to perform"* (ROMANS 4:18-21).

When the faith of God is released through you, there will be a spiritual Isaac born to you. When God's life flows through you, He will produce the impossible. Almost everybody can produce an Ishmael, a child of the flesh. This was Abraham's child through his maid Hagar; but only God can produce an Isaac, a child of the Spirit. This is the wonder of faith, that through God, who is the source of faith, we can experience supernatural and miraculous things. James taught a simple principle: *"ye have not, because ye ask not. Ye ask, and receive not, because ye ask amiss"* (JAMES 4:2-3). Many people have a difficult time asking because their faith does not see God as the source of every good and perfect gift.

3
THREE

The Third Wonder
of Faith
THE POTENTIAL
OF FAITH

For three years the primary teaching of Jesus was to build faith among His followers, to teach them the potential of faith—how to take nothing and make something happen, how to call upon God, and then watch God move. Over and over again as you watch Jesus train His disciples you see a clear emphasis towards the potential of faith that was within them. On one occasion the disciples said unto the Lord: *"Lord, Increase our faith. And the Lord said, If you had faith as a grain of mustard seed, you might say unto this sycamore tree, be thou plucked up by the root and be thou planted in the sea; and it should obey you"* (LUKE 17:5,6). What the Lord is saying in no uncertain terms is that if a bad situation is standing in your way, through faith you can pull it out and plant it in a non-obstructing area in the sea. The Gospel of Mark records the same truth: *"That whosoever shall say unto this mountain, Be thou removed and be thou cast into the sea; and shall not doubt in his heart, but shall believe that those things which he saith shall come to pass; he shall have whatsoever he saith"* (MARK 11:23) . This is limitless potential! We can have a reactional relationship of faith with God. Faith can move the impossible. A life of faith is the greatest life in this world, but it scares you to death to live it! It is not easy to demonstrate the potential of faith, for God works in mysterious ways His wonders to perform. *"For God hath not given us the spirit of fear, but of power and love and a sound mind"* (2 TIMOTHY 1:7).

Scientist tell us that there is enough power in just one little nickel to light the city of Harrisburg, Pennsylvania for a day and a night. If this is true, I don't know why our light bill is so high? Imagine the potential of power we can have to use, if we can release the power of the atoms in a nickel. I am told that man has the ability to build an automobile that can travel 100 miles per gallon of gasoline. I don't know why I am driving a car that is getting only 20 miles to a gallon of gasoline. We can become very frustrated by being told of the potential of something, and then be denied the use of it. It is not so with God. He says, "Here is my faith, it has unlimited potential to all those who believe." In the area of computers, we are amazed that a micro chip the size of a postage stamp can store all the volumes of books that are in our local library. The potential is unlimited.

God is saying to us, there is something smaller than a nickel, smaller than a micro chip, it is faith the size of a mustard seed that has enough potential to move a mountain. God has placed within us a power, the gift of the Holy Spirit, and He has given you an authority by the spoken word of faith, to destroy the enemy and see victory. That is a mighty potential. Faith connects you with God. The potential of faith is found in Jesus' statement, *"He that believeth on me, the works that I do shall he do also; and greater works than these shall he do, because I go unto my father. If you ask any thing in my name I will do it"* (JOHN 14:12-13) .

I have observed a serious problem among Christians, that they shot down the potential of great things because they did not see things instantly happening. We are accustomed to life in the fast lane. We want things now, "we can't wait for tomorrow," is the common expression of this impatient generation. Many times they shoot down the potential of great things before they happen and fail to experience the plan and will of God. There is instant pudding, instant mashed potatoes, instant chocolate cake, and even instant money via the credit card. For the Christian many times it must be an instant miracle. It doesn't always work that way; God wants first to develop His potential within you. To understand the potential of God's mighty works we must have patience. It is God at work within you. God is a miracle! This is the great wonder of faith, that the moment you begin to see God, high and mighty and sitting on a Throne of Power, in that moment things begin to happen.

4

The Fourth Wonder of Faith

THE SIGHT AND SOUNDS OF FAITH

The sight and sounds of faith produce a unique wonder. There is the voice of faith and the voice of fear. There is a voice that says, "It shall, it will, all things are possible." And then there is the voice of fear and doubt that discourages us by saying, "It cannot be, it will get worse."

Here is a biblical illustration. Jesus was with His disciples in the boat during a raging storm in Luke 8:22-25. Jesus wanted to teach them one of the great lessons of all time, the sights and sounds of faith. Jesus was asleep in the boat during this severe storm while on the sea of Galilee. The disciples, who were fisherman, heard the voice of fear, but Jesus heard the voice of faith and he was sleeping. [I am impressed with the fact that Jesus was asleep during the storm, for some of us can't sleep in a boat when it is calm. Faithless people are always jealous of those who can sleep during a storm.] During the storm the water was filling the boat and the voice of fear was saying: "Catastrophe! Catastrophe. You will drown!" Immediately they reacted to fear. They woke up Jesus screaming, "Don't you care if we perish? The sound of fear will always question God

"Where are you God when I need you."

The reason Christians are responding negatively when there is a storm in their lives, is because they are listening to the voice of fear. They are listening to the sounds that the enemy creates. Many Christians are relaxed and peaceful in the midst of a storm because they have learned to trust in the Word of God. They give the glory and praise to the sight and sounds of faith. Jesus had a sight that was different from the disciples. Jesus made a statement when He entered into the ship: *"Let us go over unto the other side of the lake"* (LUKE 8:22) . Jesus knew there was no reference in the Holy Word that the Messiah would die in a watery grave. He did read that the Messiah was going to be buried and on the third day come from the grave. Jesus knew the will and plan of God, He had the sight and sound of faith and for that reason could sleep during the storm. The disciples heard and saw the sight and sound of fear and felt hopeless. Faith can change our perspective from death to life.

The voice of faith was heard. Jesus stood up and said, *"Peace be still"* (MARK 4:39) . The winds knew the voice of faith, because it was created by that same voice in the very beginning of time. The Bible says, *"the wind and the raging of the water ceased and there was a calm"* (MARK 4:39). Fear saw the calamity, but faith saw a miracle. What do you see? What are the sounds that you are responding to? I must confess it is easy to respond to fear and negative thoughts today. We are hearing and seeing so much about gloom and doom that the voice of faith is not being heard. There is an answer and a solution to "how can we see and hear the sight and sounds of faith." *"Faith cometh by hearing and hearing by the Word of God"* (ROMANS 10:17). The great demonstrations of faith are seen in God's Word. To see is to believe. The Bible again says, *"You shall know the truth and the truth shall make you free"* (JOHN 8:32).

There is also the sound of faith on the inside of us. I am referring to the witness of the Holy Spirit. *"The Spirit itself beareth witness with our spirit, that we are the children of God"* (ROMANS 8:16). Here is the wonder of faith, that when we are in the midst of a storm, we can hear the sound of the Holy Spirit saying, *"Peace be still."* It is the spoken Word of faith. Its sight and sounds are beautiful.

The apostle Paul adds to this revelation by stating: *"Eye hath not seen, nor ear heard, neither have entered into the heart of man the things which God hath prepared for them that love Him. But God hath revealed them unto us by His Spirit: for the Spirit searcheth all things, yea, the deep things of God"* (1 CORINTHIANS 2:9-10). Through the power of the Holy Spirit within us, we can see and understand the will of God. We can know His plan for our life. *"Through faith we understand . . . "* (HEBREWS 11:3). The one thing that stabilizes us during difficult times is the sight and sound of faith. This is a great wonder.

5
FIVE

The Fifth Wonder of Faith
FAITH WORKS IN THE CRISIS HOUR

To understand this wonder I need to give a definition for the word "crisis" — "An emotional significant event in the life of a person, a decisive moment." What is a crisis? A crisis is not a dead-end street, but stepping stones to greater and better things. It is when your back is against the wall, when it appears that all the roads are blocked, or even worse, when the walls of your life are crumbling in on you and you don't know what to do. That is when faith is designed to come into action and work wonders for you. Have you been praying that you will make full use of your gifts and abilities? Do you feel paralyzed by negative feelings? Are there many unwanted questions surfacing in your life? Faith can master every crisis and make you free from the bondage of life. It is one thing to talk about faith, but it is totally different to see faith in action for the crisis hour. We need to develop the wonder of faith.

As Christians we are in a crisis hour. The actions of many Christians are hard to explain. There is a restlessness that we don't understand, a feeling of emptiness, a desire to seek but never wanting to find. We are in spiritual warfare, and the

Church of Jesus is under attack. Our Lord in Luke 18:1-8 describes, the attitude and condition of the earth just prior to His return and asks this question . *"Nevertheless when the Son of man cometh, shall he find faith on the earth?"* (LUKE 18:8) . Our faith today is being shattered. We need to see again the wonder of faith. The Bible is filled with fantastic testimonies of men and women who developed a life of faith during a crisis hour. Abraham, Joseph, Moses, David and Daniel all taught us that when there is a crisis there are priorities in a life of faith. These patriarchs of faith, by their example demonstrated to us that you don't quit when things go wrong. They encouraged us to see our problems and difficulties through the eyes of faith.

Four Steps to Develop Faith For the Crisis Hour

First, we must cultivate a life of faith. It just doesn't happen. It is a process of planting and reaping, digging and sowing, weeding and caring. Faith must be nurtured. It must be developed. You don't place faith on a shelf for if you do it will die. Here is the wonder, you can cultivate the faith of God. It is a simple operation, plant a seed and get a watermelon! When you begin to perpetuate faith it grows. It will allow you to do signs and wonders for God's glory.

Second, we must put faith into action. Faith without works is dead. We must be willing to pay the price and be willing to sacrifice to see things happen. An old Scotchman operated a small rowboat business to transport passengers across the lake. A passenger noticed that on one oar was carved "faith" and on the other "works". When he asked what those words meant, the operator dropped one oar and rowed with only "works" and they just went in circles. He then did the same with the "faith" oar and they went in circles. After the demonstration the old man picked up both oars together and sped quickly over the water. James gives us a great revelation of the truth that faith without works is dead, but works without faith is also dead. *"For as the body without the spirit is dead, so faith without works is dead also"* (JAMES 2:26). Faith leads to action. God has a special way to deal with His children. He first gives us a promise and the promise creates faith which produces an action and the action becomes the fulfillment. It will cause your feet to run, your hands to move, and your mind to think. It is an inner power that will promote an outward action. Turn faith loose within your life and watch the great things that will happen.

Third, we need to create productive goals. These patriarchs of faith enlightened us that every person should be a goal setter. The secret lies with us. We must see the need and then be open to God for His help. If we do not create productive goals, we will tie the hands of the Holy One of Israel. The God who promised to perform signs and wonders through us desires that we set a goal as recommended by Jesus. *"Ask, and it shall be given you; seek, and ye shall find; knock, and it shall be opened unto you"* (Matthew 7:7). Paul was willing to share his secret of spiritual success with the Philippian Christians. *"Brethren, I count not myself to have apprehended: but this one thing I do, forgetting those things which are behind and reaching forth unto those things which are before, I press toward the mark for the prize of the high calling of God in Christ Jesus"* (PHILIPPIANS 3:13-14). It all narrows down to one basic question. Am I going to be a hearer only or am I going to be a doer also? A doer of the Word uses faith to create goals and then applies the faith of God to make it happen.

Fourth, we need to commit ourselves to a permanent cause. One of the virtues of faith is that it stabilizes our lives. The power of faith is the power of commitment. Through faith we see ourselves in the big picture, part of God's plan and will.

We need to be committed to a permanent cause and only faith can open our eyes to see it. Our forefathers taught us by their example that our priority in life is to become part of a cause larger than ourselves. It seems easy to back away from commitment, or turn away from involvement, but when there is an eternal cause that is burning in our heart, it is easy to say "yes" to God.

What is faith for the crisis hour? It is an understanding of certain values. When you come to a decisive moment in your life and a decision must be made, you will feel the courage and strength of God. Here is our promise: *"He giveth power to the faint, and to them that have no might, He increaseth strength. Even the youth shall faint and be weary and the young men shall utterly fall. But they that wait upon the Lord shall renew their strength, they shall mount up with wings as eagles, they shall run and not be weary, and they shall walk and not faint"* (ISAIAH 40:29-31). In this promise is the assurance that we can do, even when we do not have anything by which to do it. I am aware that there are problems too complicated for us to solve, burdens too heavy for us to bear, a work too difficult for us to do, but God is able to do all things. He is the God of the crisis hour.

The Sixth Wonder
of Faith
THROUGH FAITH WE
HAVE ACCESS TO GRACE

"*By whom also we have access by faith into the grace in which we stand, and rejoice in hope of the glory of God*" (ROMANS 5:2). What is His grace? Grace is God's kindness toward us. It is God doing for us what we can't do for ourselves. I cannot save myself. When we place in proper prospective that we are the creature and He is the Creator, only then can we understand the value of His grace. Everything we have, own, or possesses is because He has allowed us by faith access to His storehouse of blessings. This is how the writer of the book of Hebrews describes it: "*Let us therefore come boldly unto the throne of grace, that we may obtain mercy, and find grace to held in time of need*" (HEBREWS 4:16). This is the great wonder of faith.

The definition of the word "grace" explains it all: Webster calls it "unmerited divine assistance given man for his regeneration and sanctification." Because of faith, at conversion we turn in our stained torn garment and God through Jesus Christ gives us a new clean garment to wear. This is how the Apostle Paul addressed the Corinthian Church: "*But God hath chosen the foolish things of the world to confound the wise; and God hath chosen the weak things of the world to confound the things which are mighty . . . But of Him are you in Christ Jesus, who of God is made unto us wisdom, and righteousness, and sanctification and redemption: that, according as it is written He that glorieth, let him glory in the Lord"* (1 CORINTHIANS 1:27-31). In Living Letters, verse 30 is translated this way: *"For it is from God alone that you have your life through Christ Jesus. He showed us God's plan of salvation; He was the one who made us acceptable to God; He made us pure and holy and gave Himself to purchase our salvation."*

This is a great wonder that through faith we accept our new position in God. In counseling sessions with people I have discovered that the number one problem Christians have is trying to do something about their past. We accept the grace of God to forgive our sins, but we limit His grace by carrying the guilt of the past. The greatest benefit I can do for people is to have a funeral in my office to bury the past! When that is allowed to happen, (and it is done by faith) you can see the change in life, personality and spirit. It is like a heavy weight taken off the shoulders of a Christian.

Are you covered by the blood of Jesus Christ? Are you a born again Christian? Are your sins in the sea of God's forgetfulness? If your answers are "Yes," then why are you worried about the past. You don't have a past! Grace is in your present! Faith allows you to see grace, and grace allows you to see your present access to God. It is because of grace that we have dominion over everything on earth. By faith we have power to loose and to bind, to remit and forgive (JOHN 20:22,23). God has given us authority so that we can say to the world, "I am what I am by the grace of God. By His grace, I am redeemed."

We have gained access to the economy of God by faith. If

we live a faithless life we will be living with constant discouragement, such as, "I can't, I'm going to fall, I'm going to backslide, I'm unworthy." But when we can see that we have access to God's grace, we discover that we are more than a conqueror through Jesus Christ our Lord. Faith is like a taxicab that takes me to the throne room of God's triumph — His victory, His provision.

SEVEN

The Seventh Wonder of Faith

FAITH ENLARGES YOUR TERRITORY

What is positive faith? Positive faith is the ability to plant a seed and watch it grow! It doesn't happen over night, but it happens. It is God's will that we "enlarge our territory." It is God's desire that we "grow and become strong in faith." Jesus taught us to pray, *"Thy will be done in earth as it is in heaven"* (MATTHEW 6:10). God has a plan and purpose for our lives. Here is the wonder of faith, we can know and understand it.

There is a biblical secret in the Pauline Epistles. When something is already atoned you don't strive to attain it. When a gift is provided, it isn't necessary to wrestle to receive it. The secret lies not in the attainment, but in the atonement. The price has already been paid. Faith says loud and clear, "It is mine!" We need to see the power of positive faith. Faith dares to move ahead, building vision into reality. Doubt cringes Faith is bold. Faith expects God to open doors and gives us strength for every task. People of faith have accomplished deeds that seemed utterly impossible. This is the challenging truth in God's Word. God desires His people to go forward, to possess the land, enlarge their borders, and be stronger today than they were yesterday.

Imagine how Gideon must have felt when God rejected all but 300 of his army of 32,000 men and then told him that he would be victorious over the Midianites (SEE JUDGES 7:1-25). As a general, this was hard for Gideon to understand, for he was taught that victory comes because of physical strength. He had to see by faith that 300 men and God were stronger and mightier than all of the armies of the world. Consider how these 300 men were armed for battle. Each man had a trumpet, a pitcher and a lamp; yet this tiny company overcame their formidable enemy by God's power. It was God's plan to win, and Gideon never lost sight of that plan. (SEE JUDGES 7:1-21). There have been thousands of battles fought through the centuries, that are forgotten, but this battle lives on as a testimony and a monument because faith was able to triumph.

Faith brings the impossible within our reach because it taps the resources of Almighty God. Doubt imprisons, but faith releases. Doubt retreats, but faith advances. Let faith widen your territory, expand your horizon, and lengthen your coastline; Then lamps of "hope" shall shine as beacons of light. Many people read about faith, but there are people who live a life of faith who believe God and trust Him. Faith is complete relaxation in the trust that what you are looking forward to has already happened.

PART THREE

THE WONDER OF HEALING

Mark 16:17-20

One of the reasons healing is not being preached today nor Christians sharing this dynamic biblical truth is that Satan has put a cloud of doubt on the subject of healing. When some people are not healed even though they have prayed and fasted, Satan attempts to shatter the confidence of Christians with doubt and unbelief. We need to see again the wonders of healing.

I must begin with this powerful statement: We are living in the end times and the Bible reveals that there will be great tribulation on the earth, sickness, disease, famine, pestilences and death (SEE MATTHEW 24:3-14). The purpose of this message is to establish within you beyond any shadow of doubt, the fact that God wants to heal you. No matter what Satan is doing on the earth, God is still the creator of the universe. No matter what Satan is doing to bring a chaotic condition physically to the human body with diseases and pestilences, I am confident that God is able to keep, protect and deliver His people. Jesus described the feelings that people will have before He returns: *"Men's hearts failing them for fear and for looking after those things which are coming on the earth"* (LUKE 21:26). There is fear that we will receive a sickness or disease by the food we eat, the air we breath, the water we drink, and by the people we associate with each day. We are all aware that the human body is under severe attack. However, God has not left us hopeless and helpless. We need to return to biblical principles if we are to understand healing.

There is a remarkable scripture in the book of Exodus that centers on an incident in the lives of the Jewish people. As they entered the wilderness they went a long period of time without water. When they came to Marah they could not drink the water because it was bitter, possibly contaminated. The people didn't know what to do so Moses called on the Lord God for help. God instructed Moses to cast a tree into the water, and the water became sweet. They drank the water without fear. The people received a directive from God: *"If thou will diligently hearken to the voice of the Lord thy God and will do that which is right in his sight, and will give ear to his commandments, and keep all his statutes, I will put none of these diseases upon thee, which I have brought upon the Egyptians; for I am the Lord that healeth thee"* (EXODUS 15:26). We can enjoy a covenant relationship with God. He will be Jehovah-rophe The Lord Who Heals, because the Cross of Jesus Christ has been applied to the bitter experiences of our life and made them sweet.

Until you are fully convinced that God wants you to be well in your body, there will always be a doubt in your mind as to whether or not you will be healed. When doubt is present, perfect health cannot exist. Healing is the result of two factors. First, the provision has been made by God through the death and resurrection of Jesus, God's perfect sacrifice for sin and disease. Second, the element of faith must be applied to the area of need, whether it is for the body, soul or spirit. Christians are fully aware of the work of redemption but are confused about faith. Some may say, "I just can't seem to get faith to be healed." Let me ask you. "Have you confidence that God will keep His promise to you?" Well that is faith! " . . . *for he that cometh to God must believe that He is, and that He is a rewarder of them that diligently seek Him"* (HEBREWS 11:6).

To set a foundational truth for healing, I would like to share a definition for healing. "The *gift of healing* is the special ability that God gives to members of the Body of Christ, to serve as human vessels through whom it pleases God to cure illness and restore health apart from the use of natural means." The gift of healing does not give a person supernatural power over disease. He or she is simply a channel through whom God works when He desires to heal. To claim that healing was only confined to the first century or just certain periods of human history, is to deny the Scriptures that *"Jesus Christ the same yesterday, and to day, and for ever"* (HEBREWS 13:8).

There is a story that has influenced my life. It was told to us at Bible College, in one of our theology classes in 1952. A first generation Rabbi went to a certain place, built a fire and said a prayer and all was well. The second generation Rabbi went to the same place and said a prayer, but couldn't build a fire. The third generation Rabbi went to the same place, couldn't build a fire and couldn't say a prayer. The fourth generation Rabbi, couldn't build a fire, couldn't say a prayer, neither could he find the place; he could only tell the story.

On such a vital subject as healing, it is sad that the best we can do as people who are empowered by God and spirit filled is to only tell the story. Just think that the majority of the Christian churches have within their tenets of faith the teaching of divine healing, but many are confined to just telling the story. It seems that we do not know how to build a fire anymore, we don't know how to pray the prayer anymore, we even forgot the place, and all we have done to satisfy ourselves is to tell the story. We need to review the wonders of healing in the light of God's Word. We must bring back the reality of healing and bring back the message of deliverance to the church and to the people that have lost hope. I heard someone say, "If God said it, I believe it, and that settles it." We need to understand things in the light of God's Word. If God said it, that settles it, whether I believe it or not. We must bring back the biblical position of healing.

The First Wonder
of Healing
PROVISIONS ARE MADE
IN THE ATONEMENT

Through the fall of man we lost everything. Our nature is depraved, our spirit is dead to God, our body is subject to disease and sickness, and our mind is carnally motivated. But Jesus recovered all that was lost in the fall through His atonement for us on the cross. There is salvation in the atonement, there is healing in the atonement, there is power in the atonement. The Christian has the right, privilege and honor to claim by faith the complete work of Jesus when He died on the cross and gave His blood as atonement. There would have been no reason for John the Baptist to introduce Jesus as *". . . the Lamb of God which taketh away the sin of the world"* (JOHN 1:29) if His death was not a sacrifice for atonement.

Isaiah understood the purpose of the coming Messiah when he prophesied. *"Surely he hath borne our griefs,* [sickness and diseases] *and carried our sorrows,* [pain] *. . . But He was wounded for our transgression, he was bruised for our iniquities: the chastisement of our peace was upon him; and with His stripes we are healed"* (ISAIAH 53:4-5). Then Peter follows the same prophetic word when he said: *"Who his own self bare our sins in His own body on* *the tree, that we, being dead to sin, should live unto righteousness: by whose stripes ye were healed"* (I PETER 2:24). Peter links the salvation of your soul to the healing of your body. Sin and sickness are Satan's twin evils, designed to tear down, to kill, and to destroy the human race, which is God's creation. Salvation from sin and healing from sickness are God's twin mercies provided to meet every spiritual and physical need of man (JOHN 10:10). It is a great comfort to know that the God who loves my soul, loves my body also.

The people who are saved from sin, and do not claim healing for their body are like the man who purchased for himself and his family a ticket to cross the ocean to go to America. They didn't have much money, so they took a big basket of cheese and crackers to eat during the trip. They boarded the ship and at mealtime every day they ate cheese and crackers. They were thrilled by the thought that they were going to America. The smell from the dining room got to them, and they asked how much it would cost to eat just one meal. They were told that when they purchased the ticket it included all their meals. What a discovery they made! The price was already paid and all they had to do was to participate in the dining hall. There is no revelation more transforming to you than to wake up to the truth that when Jesus died, He died for the healing of your body, soul and spirit. There is power in the blood of Jesus. When we lean to our own understanding, the adamic nature, which is the nature that we have been born with, provides a humanistic substitute for the needs of the inner man. This leads to disappointment and frustration. God has a better plan. Provisions to satisfy the spirit, soul and body were made possible through the atoning work of Jesus on the Cross.

The Second Wonder of Healing
IT IS FOR ALL, NO ONE IS EXCLUDED

The greatest barrier to the faith of many that are seeking bodily healing is the uncertainty in their minds as to the truth that it is the will of God to heal all. If God can heal one person, then He can heal all people. We need to correct the narrow mentality that says, "It is for someone else, but not for me." This is contrary to biblical teaching. Nearly everyone knows that God does heal some, but modern theology keeps people from knowing what the Bible teaches. God is no respecter of persons. There are no exclusions with God. Healing is not for just the inner circle, or Christians that have been following Jesus for many years. Here is the wonder of healing, that even if you are a young Christian, or a child, or have little knowledge about theology, everyone has the same privilege and honor to come to the throne of grace boldly and present their petition for healing. The door is open to everyone, the weak and the strong. Everyone has the same footing in the area of healing. "Lord God heal me."

It is difficult to boldly claim healing when you have questions as to whether or not healing is for us today. There are

Christians who would rather hear the opinions of man than to search the Word of God. The power of God can only be claimed where the will of God is known. It would be next to impossible to get a sinner to *"believe unto righteousness"* before you had fully convinced him that it is God's will to save him. Faith begins, when the will of God is known, when you are aware of what God can do. I have witnessed many people who could have been healed, but were not because they felt unworthy, or inadequate to come to God. There is a strong temptation to share your prayer request with someone, yet in your heart feel that God won't hear your request. *"Be careful for nothing; but in every thing by prayer and supplication with thanksgiving let your requests be made known unto God"* (PHILIPPIANS 4:6).

When the leper appealed to Jesus for healing, he said, *"Lord, if thou wilt, thou canst make me clean"* (MATTHEW 8:2). He was confident of the ability of Christ, but uncertain of His willingness. Jesus immediately corrected his mistaken concept. *"I will; be thou clean"* (MATTHEW 8:3). Christ's *"I will"* canceled the leper's *"if"* and added to his faith that He could be healed. The missing key in the leper's life was the knowledge that Christ desired to heal him. You must take the "if" out of your life. The community degraded him to a second class citizen, an outcast, but he found that when he came to God, he was a first class citizen. All of us can come to God as we are with full assurance that He will not reject us.

The secret of receiving from God is the simple truth that when you see that your name can be added to the promise, your faith is made strong. When I accepted Jesus as my Savior, the soulwinner made me read John 3:16 and wanted me to hear my name in that Scripture. *"For God so loved, [Al Alaimo] that He gave His only begotten Son, that whosoever believeth on Him should not perish but have everlasting life."* When I realized I was personally part of God's plan for salvation, it was easy for me to claim conversion for my life. When I read God's promise to heal and I can place my name there and I realize that the will of God is made known, then I can claim it. My faith will be made strong because God has not excluded me. That is a great wonder. There is a promise in 3 John 2 where I can place my name in Scripture for my healing. *"Beloved* [Al Alaimo], *I wish above all things that thou mayest prosper and be in health, even as thy soul prospereth."* Do not allow things or circumstances to keep you from coming to God and asking Him to heal you.

THREE

The Third Wonder
of Healing
GOD'S PURPOSE
IS REVEALED

The right to pray and receive answers from God is our privilege. In Luke 17:5 the apostles asked Jesus: *"Lord, increase our faith."* The Bible states that Timothy was *"nourished up in the words of faith"* (1 TIMOTHY 4:6). In Jude 3 the Word of God declares: *"earnestly contend for the faith."* The Word of God is the prescription for receiving faith. *"Faith cometh by hearing and hearing by the Word of God"* (ROMANS 10:17). Don't pray for faith to be healed; have faith in God. Hear the Word. As you learn to know God's Word, you will have faith. The Word of God develops faith.

There is an interesting lesson in Mark 9:14-29: A father came to Jesus with a demoniac son. Jesus said to the father, *"If thou canst believe, all things are possible to him that believeth."* The father cried out, *"Lord, I believe; help thou mine unbelief"* (MARK 9:23,24). Unbelief is the inability to understand the purpose of God. It is one thing to know that you are not excluded from God's plan to heal, but it is something different when you read God's Word and see how the plan of God is revealed. A doubter often prays for things already available to him.

God's purpose to heal is revealed through people who are delivered from sickness and when we see the works of the devil destroyed. The leaders of the early church, like Peter, kept the message of healing alive. *"The word which God sent unto the children of Israel preaching peace by Jesus Christ: that word, I say, ye know, which was published throughout all Judea, and began from Galilee, after the baptism which John preached; How God anointed Jesus of Nazareth with the Holy Spirit and with power, who sent about doing good, and healing all that were oppressed of the devil; for God was with him"* (ACTS 10:36-38). So it is today. This ministry is continuing as promised, because it is the same Word that comes from God. *"Greater works than these shall ye do,"* was the promise that Jesus gave us (JOHN 14:12). Allow God to reveal to you that because Jesus is *"touched with the feeling of our infirmities"* (HEBREWS 4:15) and because He knows He cares, and because He cares, you can be healed. The psalmist David understood God's purpose revealed in healing! *"Then they cried unto the Lord in their trouble and he saveth them out of their distresses. He sent His word, and healed them and delivered them from their destruction"* (PSALM 107:19-20). Oh the wonder of His purpose revealed in healing!

God's purpose to heal is revealed to establish Jesus' astonishing claims. Can we believe His promises? How can men be convinced of healing? The claims of Jesus were not only for the first century. *"For the promise is unto you, and to your children, and to all that are afar off, even as many as the Lord our God shall call"* (ACTS 2:39). The claim of Jesus was the truth that there was nothing too hard for God. God is the God of the impossible. The Bible tells us why things do not happen: *"ye have not, because ye ask not"* (JAMES 4:2). Jesus gives us an example of how the power of God works. There stood before Him a paralyzed body and a sinful soul. Which would be easier, to say, *"Thy sins be forgiven,"* or *"Take up thy bed and walk?"* (Mark 2:9). Jesus made it clear that the power to work the one miracle involves the same power to work the other. For Jesus said, it is no easier to redeem a life from sin than it is to heal his body from diseases. The reason why we are not seeing more healing is because the astonishing claims of Jesus are being minimized. I have observed that when a church reveals the need of evangelism, souls are saved. When a church reveals the need for discipleship, there is a wide concentration on teaching. When a church reveals that the heart of God reaches out to the sick, hurting, and troubled lives, there will be a great demonstration of God's power to heal and deliver. The Church must not be silent with the message of divine healing.

God's purpose to heal is revealed to empower the Gospel message as preached by His people. *"These signs shall follow them that believe . . . they shall lay hands on the sick, and they shall recover"* (MARK 16:17,18). There is never a great deliverance without a great sickness. There is a very personal message from God. He will take care of you. If you can't believe that, then you have not understood the wonder of the purpose revealed in healing. The major theme of the apostle Paul was that the mystery of the ages has been revealed. *"To whom God would make known what is the riches of the glory of this mystery among the Gentiles; which is, Christ in you the hope of glory"* (COLOSSIANS 1:27). It is because Jesus Christ lives within us that we can claim *"if the Spirit that raised up Jesus from the dead dwell in you, He that raised up Christ from the dead shall also quicken your mortal bodies by His Spirit that dwelleth in you"* (ROMANS 8:11).

God's purpose to heal is revealed to establish the resurrection of Jesus. The resurrection of Jesus is to validate that there is power to resurrect the physical body from weakness to strength, from sickness to health. The whole physical and spiritual concept of the resurrection of Jesus is the miracle that God did the impossible. If you can believe in the resurrection then you can believe that God can make the impossible, possible.

God's purpose to heal is revealed to bring glory to His kingdom, to inspire the faith of believers, and to encourage the ones who are seeking healing. It is true, everybody wants to be healed, but we must examine our hearts. Will God get maximum glory through your healing? God's purpose in healing you is that the message of healing will continue to grow because of your continual praise to Him. What you have endured through your affliction is to bless someone and to bless the kingdom of God. Our prayer should be, "Lord, if I must go through this road of affliction, then I want to be a witness of your glory, power and might." Satan meant it for evil, God turned it for good. We should have a healing testimony so that we can influence the world.

FOUR

The Fourth Wonder of Healing
PROVISION IS ESTABLISHED IN THE CHURCH

"*Is any among you afflicted? Let him pray. Is any merry? Let him sing psalms. Is any sick among you? Let him call for the elders of the church; and let them pray over him, anointing him with oil in the name of the Lord. And the prayer of faith shall save the sick, and the Lord shall raise him up; and if he has committed sins, they shall be forgiven*" (JAMES 5:13-15). Concerning the "prayer of faith," many people have the idea that this means an immediate answer, or instant results. This may be true, but if not, we should not be become discouraged. Once the prayer of faith is offered and the disease is rebuked, the case is in the hands of the Lord and He does the restoring. Whether healing is done instantly or is gradual, we should trust God, and stand on His Word.

We have violated this ministry in a very serious way, because we have judged healing on what happens three minutes after someone has laid their hands on our head. We have done an injustice to the plan of God. This is a very sacred doctrine. We can discover the wonder of healing in it. We need to come back to the simplistic plan God gave to the church. We need to call for the pastor and elders, which are the spiritual leaders of the Church. *"Is any sick among you, let him call . . . "* The responsibility is placed on the afflicted to come believing that God can. Spiritual leadership has the responsibility to be obedient to God's Word by anointing with oil and praying the prayer of faith. It is a prayer that believes that God can do the impossible. Here is the wonder of healing, two people agreeing, the afflicted and the believer, that the Lord shall raise him up. I have witnessed hundreds of healing through this simple but powerful plan of God.

I have asked God on numerous occasions why more people aren't healed after we have followed this scriptural pattern. The Holy Spirit called to my attention that there is an important area in healing that has been ignored; *"If he has committed sins, they shall be forgiven."* There should always be a heart searching experience before we ask God to heal us. God also wants to heal the sin in your life. In Bible days Jesus healed sinners and said *"go, and sin no more"* (JOHN 8:11). God makes the provision to forgive sin and set you free from condemnation and bondage. When we honor God's Word, God will honor us. Why does Christ heal? Because of compassion. This is the reason why He came to this earth to die. *"God commended His love towards us in that while we were yet sinners, Christ died for us"* (ROMANS 5:8).

Healing is the children's bread. Our Father is like the earthly father who stands his little son on a step and kneels before him inviting the boy to jump. "Son, jump to Daddy." The little boy is naturally frightened at the monstrous distance of the step and the sidewalk. "C'mon Son! I'll catch you." Finally, taking a deep breath, he plunges into the arms of his Father. In that jump, a young mind has learned a great lesson in trust. He didn't learn trust standing on the step, He learned trust when he jumped.

God wants to teach us a lesson in trust when He says to the afflicted, "Jump! Trust Me! Believe My Word! I am the Lord that healeth thee." Christ has placed this ministry in the church and He wants us to practice it, to believe it, He wants you to plunge in it and not to be afraid. His arms are there to catch you.

FIVE

The Fifth Wonder
of Healing

DIVINE POWER IS RELEASED
THROUGH HUMANS

Genesis 1 tells how God created the world. We see by the Word of God what was brought into existence. When God said, *"Let there be . . . "* it happened. We need to understand the creative power of His Word. God's Word contains what God says. He will make possible the impossible. Jesus said, *" The **words** that I **speak** , they are **spirit**, and they are life"* (JOHN 6:63). God has given us His Word. We must speak His Word. We have the ability, power and privilege as a human being, to speak the Word of our Divine Creator. That is a great wonder. We can become a channel through whom God can flow His mighty power to heal and do miracles.

It has been said, "The influence of what is said depends largely on who said it." When God speaks, all the creative power that was in the beginning when He spoke the world into existence again goes into action. If God could create a world with three words, *"Let there be"*, then surely He can heal your sick body with His Word. His words have power. The Bible assures us, *"He sent His Word and healed them"* (PSALM 107:20). God's Word is with power when accepted and acted upon in faith. God named Jesus "the Word." *"In the beginning was the Word and the Word was with God, and the Word was God"* (JOHN 1:1). The miracle in the belief of healing is that He has given us His Word. God has given us, we who are unworthy human beings, authority in the name of Jesus. We can speak His Word, we can claim the power of His Word, and it is released to us through the name of Jesus.

There is something we can physically do to be healed. Write the promises of the Word and keep them visually before you. Learn to digest the Word within your sick body. I have encouraged people to write promises of God on a 3x5 card and keep it visible and read it at least seven times a day. As a human being we do not have the power to heal ourselves or others, but we do have the power to apply the One who can, He is Jesus, the Word. The Word has the power. The Word is what you will feel in the midnight hour when the pain is too difficult to bear. You can feel the over-shadowing presence of God. That is what Jesus was revealing when He said, *"The words that I speak unto you"* are not just ordinary words, *"they are spirit , and they are life"* (JOHN 6:63). If that were true then there is something I can do; I can allow this divine power to be released through my life.

When Mary was told that she would conceive a child by the Holy Spirit and that He would save His people from their sins, she understood that in the natural it was impossible. There was the frightening thought of reproach, misunderstanding, pain and even rejection within the heart and life of Mary. Mary gave us the secret of favor with God when she said: *"Be it unto me according to thy Word"* (LUKE 1:38). That was what God wanted to hear.

You and I have the Word. Take it! Use it! Don't abuse it! Let the Word be stamped on your heart. Let the Word be visible. The power is in the Word revealed to us, and it flows through us. There is an alarming caution I would like to share. There are people that will travel hundreds of miles to be in the presence of someone who has the gift of healing, but wouldn't move ten steps to pick up the Bible and read *"the Word."* Recently one of the members of our congregation went with a group of people to a major city to hear an outstanding evangelist and to witness the healing power of God. She felt that she was healed, but a second report revealed she still needed an operation. She then said something to me that is scriptural. "Pastor can you give me a list of scriptures that I can read, so that I can hold on to the Word of God." That was the greatest thing I could have even done for someone who needs healing. It released to her the power of the Word.

There is a truth that has been handed down to me since I was a young Christian. If you want to be healed, get serious with God. He desires to release His power through you, but the channels need to be open. The Word opens the channels.

6
SIX

The Sixth Wonder
of Healing
WE CAN CLAIM HEALING
BEFORE WE NEED HEALING

F.F. Bosworth in his book, *Christ the Healer* opens up a great truth. Before attempting to exercise faith for healing, one needs to know what the Scriptures plainly teach. Jesus said, *"The Word is the seed"* (SEE MARK 4:3-15). It is the seed of divine life. The person seeking healing needs to know from God's Word that it is God's will to heal.

To receive healing is not like a magic wand placed over you. It is not just being at the right place at the right time. Healing is the result of God's seed placed in your heart. The Bible teaches us that *"faith comes from hearing and hearing by the Word of God"* (ROMANS 10:17). It is impossible to receive spiritual things, if your heart and life are carnal. It is important for a Christian to *"grow in grace and in the knowledge of our Lord and Savior"* (2 PETER 3:18). The Christian life is a process of learning, growing, developing, so that when the hour of need comes to us, the spirit of the Lord will come alive within us. Isaiah saw the same picture. *"When the enemy shall come in like a flood, the Spirit of the LORD shall lift up a standard against him"* (ISAIAH 59:19). Until the person seeking healing is sure from God's Word that it is God's will to heal him, he is trying to reap a harvest where there is no seed planted.

One of the vital keys to a life of healing is a life that is yielding a spiritual harvest. Jesus said, *"Ye shall know the truth and the truth shall set you free"* (JOHN 8:32). Freedom from sickness comes from knowing the truth. We need to hear more of the importance of living close to God, to be led by the Spirit, and to develop a Spirit awareness in our lives. It is because we know God's Word and we are listening to the still small voice that we claim His protection, even before we are engulfed with sickness.

God is the healer. When God is in us, then healing is in us, protection from sickness is in us The greatest power in all the world is in us. The awareness of healing is a way of life for the Christian. In the morning we ask God to protect us and our family and cover us with the precious blood of Jesus. Throughout the day we feel a hedge of protection around us. Every time we eat a meal, we ask for the Lord to protect us from harm. We close the day, thanking our Lord for watching over us. This daily practice has the power of healing in it—a healing before we need healing. This is a wonder to me that we can sow a seed, pray a prayer, claim a protection, even before the need of healing.

Observe how Jesus taught us to pray in Matthew 6:9-13. We are to ask for bread before we are hungry. *"Give us this day our daily bread."* Lead us away from temptation before temptation comes. *"Lead us not into temptation."* There is a biblical principle that Jesus taught. *"If the good man of the house had known in what watch the thief would come he would have watched and would not have suffered his house to be broken up"* (MATTHEW 24:43). I believe it is important for us to seek the provision of healing for our lives every day. Place a hedge of protection around your family every morning. At each meal ask God to bless and sanctify the food so that you can remain in perfect health. We should claim healing even before there is a need for healing. How do we prepared for healing? —by daily praising God for being our healer. There are seven redemptive names of God, one of the names is Jehovah-raphe, *"I am the Lord that healeth thee"* (EXODUS 15:26). "I am the Lord thy Physician." Isn't it great to know a physician before you get sick. Nothing is more frightening than to be in a foreign country and be rushed to the hospital and when they ask for a doctor's name, you respond, "I don't know one." God's people have a physician, He is One they believe in, He is Jehovah-raphe, the Lord my Physician. This was the first covenant after crossing the Red Sea on their way to the Promise Land. This is not only a promise; it is a statute and an ordinance. This is a great wonder to me, because we do not have to go out in the 11th hour seeking healing. We have had the Healer by our side all of our lifetime. We have been claiming healing before we needed healing.

SEVEN

The Seventh Wonder of Healing
HEALING OPERATES THROUGH FAITH

What is faith? *"Now faith is the substance of things hoped for, the evidence of things not seen"* (HEBREWS 11:1). This verse is sometimes quoted: *"Faith is the title deed for things you have hoped for, the putting to proof of things unseen."* Faith is the title deed to the property you know you possess, even though you have not yet seen it. This is a wonder that through faith you can see your healing long before your physical eyes can see it happen. This is not mind over matter. This is not talking yourself into some pipe dream. This is a relationship of faith that you have with your God.

God has never asked that we exercise faith for something that He has not first promised to do for us. Our eyes are not on the affliction, or the sickness, or the disease, our eyes of faith are focused on God, who is *"the author and finisher of our faith"* (HEBREWS 12:2). We release the burden to Him and He says *"My grace is sufficient for thee: for my strength is made perfect in weakness."* (2 CORINTHIANS 12:9). We live in perfect assurance that God will make a way. He knows a way through the wilderness and all we have to do is follow.

This is how God deals with His children: First, He gives us a promise, and when the promise creates faith, faith produces an action, and in that action, He fulfills the promise. We are in spiritual warfare today, and the battle lines are drawn. Satan is attempting to destroy our faith. *"God hath not given us the spirit of fear, but of power and love and a sound mind"* (1 TIMOTHY 1:7). If we yield to the spirit of this world, we will be filled with fear and doubt. We cannot afford to relinquish our faith to the enemy. We cannot allow our faith to die. Faith is the element that moves the hand of God. Keep your faith strong!

Balaam, a true prophet of the Lord said, *"God is not a man, that He should lie; neither the son of man, that He should repent; hath He said and shall He not do it? or hath He spoken, and shall He not make it good?"* (NUMBERS 23:19). The prophet was really saying. Besides God, who else can you trust? Man lies, but God never lies. Look at the record of God and how He lives up to what He says. Trust God, He will never fail.

Now that you know your rights in Christ as a believer, act on your faith. Rebuke the enemy that has interfered with your health and be benefited according to the promise. Do as Abraham did. He waxed strong in faith by looking unto the promise. *"Abraham staggered not at the promise of God through unbelief; but was strong in faith, giving glory to God; and being fully persuaded that, what He had promised, He was able also to perform"* (ROMANS 4:20,21). Start today to live in the promise. Say it and believe it, "The Lord is my healer." The wonder of healing is . . . that He healed me!

PART FOUR

THE WONDER
OF PRAYER

Psalm 6:1-2

D avid had the proper concept concerning prayer. During recent months we have sought guidance from God's Word on the discipline of prayer. The greatest lessons we learn in life are learned by example. Jesus is that perfect example on the subject of prayer. The disciples watched and listened to Jesus pray and as a result felt their own shortcoming concerning prayer. They had an honest and sincere request, *"Lord, teach us to pray"* (LUKE 11:1). They had seen and witnessed the mighty power of miracles, but they never asked for power to do miracles. They had heard the matchless preaching and teaching of Jesus, but they never asked Jesus to teach them how to preach to the multitudes. They longed for only one thing, His dynamic prayer life. You cannot read the life story of Jesus without having a longing in your heart to share the same ministry and experiences of a prayer life. The first observation we make concerning the prayer life of Jesus was the request by the disciples. It was not what to pray, but *to pray.*

It has been said that prayer is the breath of the new man, the motion of the renewed heart, and is essential to the existence of the Christian life. Prayer is not simply an act, it is an attitude. Prayer is not simply a ritual, it is a relationship. Prayer cannot be regulated or legislated. Prayer is a way of life. Prayer is an experience you have with God in the Spirit. There is need today to see the "wonders of prayer." We need to revive the ministry of prayer in our lives, it is under extreme attack. People are not praying as they should, for they would rather light a candle, or make a phone call to some prayer center, or they would rather put their request on a prayer chain. All of these activities may be proper and fine, but God has a desire that prayer will be developed in our personal lives. We have lost the wonder of the ministry of prayer, and it must come back if we are to survive in the next decade.

The spiritual poverty that plagues our lives and the cold orthodoxy that structures our theology must give way to the dynamics of the Holy Spirit's involvement of prayer within our lives. Effective prayer is found only in the feeling of absolute necessity and in the faith that recognizes God's will and identifies with it.

Prayer is not overcoming God's silence, it is laying hold of God's willingness. To pray is to turn the request over to God and say, "God, you are in control, you are sovereign, you have the power to do all things, and I seek your will." What is the principle of prayer? Prayer is not talking God into doing something He doesn't want to do, but rather it is giving Him the moral right to do what He has been wanting to do for a long time.

All of life has its own laws. Where these laws are followed, life is sound and strong and easily lived and it will bear an abundance of fruit. If we break these laws, life will be burdensome and fruitless. The prayer life has its laws! If we pray contrary to the will and desire of God, if we pray according to our own will, then prayer will be meaningless.

There are several warning about prayer that we need to understand before we can experience the wonders of prayer. Some Christians have stopped praying, others are confused about prayer, and others are just misguided with the belief that if you say it, you've got it.

Some Warnings To Guide You.

First, we think that we must help God to fulfill our prayer. We are commanded only to pray. It is God Himself that will take care of answering and fulfilling. God is very capable of doing things when and how He feels best for us and His kingdom. You don't have to help God do it; all you have to do is pray and leave the results to God. If answers don't come when you think they should come, do not charge God or get discouraged, just keep on praying.

Second, we sometimes use prayer for the purpose of commanding God to do our bidding. God has not given us His promises in order that we will make demands or compel Him to do what we ask. God has not given us His Word so that it would be used against Him. When we pray, we need to understand we are not commanding God: we are not ordering God, we are not compelling God. He is in total control. It is a privilege to come to Him.

Third, we forget to pray in the name of Jesus. In our prayer life nothing means as much to us as to pray in the name of Jesus. If we fail to do this, our prayer life will either die from discouragement, or despair, or become simply a duty which we feel we must perform. The name of Jesus is not just a name we feel we must use, but it is a name that God wants us to use. There is a mediator between God and man, and He is Jesus. Prayer is coming to God through the name and power of Jesus Christ.

Without prayer you will only grow to a certain capacity, but through prayer you can make the impossible happen. It all comes through the simple, the easy, the most precious experience of life; it is prayer! The Christian has lost the art of praying. It is not part of our way of life anymore. This must change again! Develop the ministry of prayer. The psalmist David said several times: *"In the day of my trouble I will call upon thee: for thou wilt answer me"* (PSALM 86:7). Not only during a crisis, not just periodically when we feel we want something, but David said *"Because he hath inclined his ear unto me, therefore will I call upon him as long as I live"* (PSALM 116:2).

We can never fully understand the ministry of prayer, unless we look at the example of Jesus. Watching and observing Jesus, whether it is in the morning, or noonday, or evening, or even all night, Jesus prayed. The Son of God prayed hours and sometimes He prayed all night. If our Lord prayed, why is it so difficult for us to see the importance of prayer in our personal lives also. Jesus did not have to pray, He was the Son of God, but He wanted to teach us that as a human we can have supernatural power in our lives when we pray. Prayer establishes a relationship with the Almighty God. When you talk to God the heavens will open and something supernatural will happen, not because of what you prayed but how you prayed. Jesus said, *"After this manner pray ye..."* (MATTHEW 6:9). He laid before them the dynamic life of prayer.

The will of God cannot be done if we do not spend time with God in prayer. How can we understand the purpose of the kingdom or the things that God desires to be done if we are not in communication with Him in prayer? God will reveal His mysteries of the kingdom to us, when we are in prayer. God can reveal to you in just a few moments of prayer what flesh and blood never reveal. Prayer changes things.

ONE

The First Wonder of Prayer

IT DEFEATS OUR SPIRITUAL FOE

Our authority over Satan is based upon the redemptive work of Jesus Christ when He died on the cross. Jesus gives us the secret of how we can regain our stolen possessions from Satan. *"If I drive out demons by the Spirit of God, then the kingdom of God has come upon you. Or again, how can anyone enter a strong man's house and carry off his possessions unless he first ties up the strong man? Then he can rob his house"* (MATTHEW 12:28-29 NIV). The strong man to whom Jesus is referring is Satan. It is time for us to rob Satan, to take back from Satan what he has originally taken from us. The possessions, which are to be taken from Satan, are the people who are enslaved by him. Does Jesus expect us to rob Satan and regain our possessions? Yes! Do we really have an authority to bind Satan? You must go into the strong man's house to take back your possessions! You must bind him. Why? He has your possessions. He has your health in his hands, he has your children in his hands, he has your future and destiny in his hands, he has your prosperity in his hand, and you must go into his house and bind him! You have a power to defeat him!

The Bible recognizes Satan as a strong man, but the Bible also recognizes that a Christian who prays is a stronger person. You and I can't walk alone into the presence of the enemy and say, "Give me back my joy, give me back my peace, give me back my purpose for living." He will laugh at us. But through the ministry of prayer, and the power of Jesus, and our relationship with God, we can tie up the strong man and rob him. Prayer brings an anointing and an authority over the forces of darkness. Jesus explains this principle of binding and loosing. *"Upon this rock I will build my church; and the gates of hell shall not prevail against it. And I will give unto thee the keys of the kingdom of heaven: and whatsoever thou shalt bind on earth shall be bound in heaven: and whatsoever thou shalt loose on earth shall be loosed in heaven"* (MATTHEW 16:19-19). How do we bind Satan? This is a wonder! Since prayer is warfare, and Satan is so powerful, we are urged by the Spirit *"to put on the full armor of God,"* and to use all the weapons God has provided for us to win our victory. The purpose of the armor is, *"so that you can take your stand against the devil's schemes"* (EPHESIANS 6:11 NIV). You may have to spend time in prayer until you feel the inner witness of the Spirit, so that you will have authority over Satan. Through prayer you have the will and spirit to win. I offer you some suggestions on what to do to defeat your enemy:

Make a positive claim to the redemptive work of the blood of Jesus Christ. You have no merit of your own. The blood of the Son of God is now available as the basis of our plea for God's answer in prayer. We must come back to this powerful truth, that there is wonder working power in the blood. When we pray we have a glimpse of Calvary, for the blod has never lost its power.

Make a positive claim through the name of Jesus. For God has given us a name that is above every name. Why is the name of Jesus powerful? Mary and Joseph did not give Jesus his name. The Bible says,*"Thou shall call His name Jesus"* (MATTHEW 1:21). It was a name given by God. That is why Satan trembles at the name of Jesus. Use the name of Jesus to bind Satan.

Make a positive claim to the sovereign authority and almighty power of Jesus our Lord. He is seated at the right hand of God. He is alive! He reigns! The devil hates that type of celebration. We should not allow Satan to discourage us with doubt because certain things have not happened, or things have not worked out according to our plans. Rejoice in the sovereignty of God!

Make a positive claim in the victory Christ has established over Satan and his demons. We rejoice in the truth that Jesus has purchased redemption with His blood, and reigns supreme over our enemy. Because He lives, we live. Because He reigns, we reign. We need to strenghten our prayer by reading scriptures that proclaim victory, such as in LUKE 10:19, JOHN 12:31, ROMANS 16:20, I CORINTHIANS 2:6, EPHESIANS 1:21, COLOSSIANS 2:15, AND HEBREWS 2:14. Praise God for Christ's complete victory!

March forward, praising God. Through prayer, praise is not only a method of worshiping the Lord, but praise is a mighty spiritual weapon. Praising God binds the enemy.

The Second Wonder
of Prayer
YOU WILL EXPERIENCE
GOD'S PEACE

"*Be careful for nothing; but in every thing by prayer and supplication with thanksgiving let your requests be made known unto God. And the peace of God, which passeth all understanding, shall keep your hearts and minds through Christ Jesus*" (PHILIPPIANS 4:6-7). There is no blessing that bring more positive benefits to the life of a Christian than the enjoyment of God's peace. While all Christians have experienced peace with God by virtue of exercising the gift of saving grace, many know little of the *"peace of God which passeth all understanding,"* because their lives are not dominated by the belief of prayer. It is in prayer that we wrestle with the "Why's" of life, or the discouraging reports that we received. To me this is a wonder. Though I don't know how it happens, God's presence overshadows me, the Spirit bears witness within me that *"all things work together for good"* (ROMANS 8:28). and He promises me that *"My grace is sufficient for thee: for my strength is made perfect in weakness"* (2 CORINTHIANS 12:9).

In prayer I get that overwhelming feeling, *"The Lord is my shepherd I shall not want"* (Psalm 23:1). There is no experience on this earth that can make you feel more secure and peaceful than the experience of prayer. We get peace that passeth all understanding. This peace goes beyond all physical reasoning: it gives you peace even when your mind says, "You can't have it." It gives you peace when your logic says, "It won't work out." There is something that happens to you when you pray. *"Men ought always to pray, and not to faint"* (Luke 18:1). It is impossible to enjoy God's peace and not face a war. The fact that we have found peace, indicates that we were in a conflict, a battle, a storm, but the presence of God delivered us, and now we have peace. The experience of peace in the midst of the war becomes the reward of victory for the Christian who prays.

The unconverted person knows nothing of peace with God or the peace of God. The Bible describes their condition: *"The wicked are like the troubled sea, when it cannot rest, whose waters cast up mire and dirt. There is no peace, saith my God, to the wicked"* (ISAIAH 57:20-21). The anxiety that paralyzes the world need not affect the children of God. God desires to guard our heart and mind against all the turmoil the world produces, and He does it through prayer. The severe storms may be on the outside of us, but the Christian must not allow it to get within his heart.

God's peace is a miracle. It is the invincible garrison that God throws around the human heart of a Christian to keep it intact against the enemy. Oh, the wonder of prayer! It not only settles the storm, but we discover the deep tranquillity of His divine presence. Are you perplexed, troubled, worried? Let God's peace guard your heart. As you enter into prayer you will feel yourself in the palm of His Hand.

Here is the wonder! How, because by prayer the Christian is overshadowed by the peace of God. When we come to Him in prayer, something is removed, and then something is replaced that gives us peace. We surrender to God our trouble, our distress, our storm; when we release it to Him in prayer, and He gives us His peace. *"Be careful* [anxious] *for nothing; but in every thing by prayer and supplication with thanksgiving let your requests be made known unto God. And the peace of God, which passeth all understanding, shall keep your hearts and minds through Jesus Christ"* (PHILIPPIANS 4:6-7). Our God is in total control. *"If the Son therefore shall make you free, you shall be free indeed"* (JOHN 8:36). That is the wonder of wonders, *"Thou wilt keep him in perfect peace, whose mind is stayed on thee: because he trusteth in thee"* (ISAIAH 26:3).

The hallmark of our generation is marked with stress, gloom and doom. It is staggering how much money is spent on tranquilizers and drugs to ease the tension and problems of life. The amount is in the billions of dollars spent annually to pay for the psychologist and counseling sessions that people feel that they need just to survive daily. The solution for our problems may seem simplistic, but I believe that it is found in the power of prayer. If a person would spend at least 15 minutes every day with God in prayer, mediating in His presence, they would find an overwhelming feeling of God's peace for trouble times. All your problems may not be solved, but you will feel strong, because *"His grace will be sufficient for you, His strength is made perfect in weakness"* (1 CORINTHIANS 12:9).

How does prayer secure the presence of God? *"For he shall give his angels charge over thee, to keep thee in all thy ways"* (PSALM 91:11). Through prayer I am aware of the presence of God. I know He will protect me. The Lord has made a way in the wilderness and all I have to do is follow. The Christian who prays has a peace, he believes God is in control: he knows God is working things out for his good, and he feels the comfort of the Holy Spirit. Develop a quiet time with God, and He will give you His peace.

THREE

The Third Wonder of Prayer

IT FORMS A BOND BETWEEN CHRIST AND THE CHURCH

Prayer brings you into a sacred partnership with Jesus Christ. If God had not revealed it in His Word, it would have been blasphemous to say it. We are bonded together with Jesus in prayer! Prayer gives us a union that unites us with Jesus. *"And they went forthand preached everywhere, the Lord working with them, and confirming the word with signs following"* (MARK 16:20). There are many ways to work with God, through obedience, through services to others, through sharing His love. But he wants a more intimate contact with us—*the Lord working with them.* He wants to bring us into His inner circle where we can hear His great heartbeat for a lost world. God created us with the ability to speak to Him and fellowship with Him. We were created to pray, as He prayed.

What an honor you have through prayer, that God the Almighty, God the Omnipotent, God the Creator, will come down to where you are and He will plan and work with you. If you would spend premium time with Him, He will develop your lifethe way it ought to be. That is the highest honor a human can ever experience on this earth. Why people do not pray more will always be a great mystery to me. God wants to work wonders and signs in and through you.

God wants a more intimate contact with us, it is called "prayer." God desires to bring us into His inner circle, so that we will feel the heartthrob of people's needs. He wants to bring us into a relationship with Him, that both we and God can agree for supernatural happenings on earth. God created us with the ability to speak to Him. No other creature on this earth has the ability to speak to God. He also created us with the ability that He could speak to us. That to me is the wonder of prayer. We were created to pray. No other creature was created to pray. Why? So that a bond will be formed with the Almighty God. Jesus said: *"And I will give unto thee the keys of the kingdom of heaven: and whatsoever thou shalt bind on earth shall be bound in heaven: and whatsoever thou shalt loose on earth shall be loosed in heaven"* (MATTHEW 16:19) .

Jesus at the present time reigns as an intercessor. He is the Great High Priest and desires us to intercede with Him. *"But ye are a chosen generation, a royal priesthood, a holy nation, a peculiar people ; that ye should show forth the praises of him who hathcalled you out of darkness into his marvelous light"* (I PETER 2:9). He desires for you also to be a *"royal priesthood."* Every time you pray, you become a partner with Jesus. The Scriptures are clear that *"Jesus is standing at the right hand of God"* (ACTS 7:55) and *"There is one God and one mediator between God and men, the man Christ Jesus"* (1 TIMOTHY 2:5).

God desires that this "golden triangle" will be the key of deliverance for our generation. What kind of prayer partner would you be if you kept disagreeing with Him? We need to know the will of God. We need to have a full revelation of God's kingdom so that we can stand in agreement with Jesus Christ our Lord. If Jesus and you are praying together about a certain matter, there should be no doubt that God is listening. We come to God in the name of Jesus, and as a blood-washed child of God. There is no possible way, God will turn a deaf ear to us. God's Word stacks assurance on assurance, to encourage your intercession and prayer life.

Jesus Himself urges you and me to hold on in prayer as an intercessor and not to give up until the answer comes. Jesus gave us many amazing stories to illustrate this truth. He encourages us to keep on knocking, like as the man who needed bread during the midnight hour. It was because of his importunity that the door was opened and he was given the bread he needed (LUKE 11:5-10). Just look how many were healed and received their miracle. You must ask in order to receive. You must seek to find. You must knock if anything is going to open to you. Here is the principle that Jesus taught: You must be persistent. Don't give up, keep on believing and you will receive. That is the spirit of one who prays. That is the heart of an intercessor. My friend, He may not come when you want Him to come. He may appear to be silent to your request. But you can mark this truth: He will come right on time according to His schedule. Remember this, you have the attention of God because of Jesus. This to me is a wonder!

We need to understand how the principle of prayer operates. God will only answer if you present a need. The key is to ask, to make your request known, to come before the throne of grace. This places the obligation on us. We can tie the hands of the Holy One of Israel, our God, because of our failure to unite with Jesus in prayer for our needs or the needs of others. If we neglect to pray, there will be no answer to the need. Jesus said, *"Verily Verily I say unto you, He that believeth on me, the works that I do shall he do also; and greater works than these shall he do; because I go unto my Father. And whatsoever ye shall ask in my name, that will I do, that the Father may be glorified in the Son. If ye shall ask anything in my name, I will do it"* (JOHN 14:12-14). There is a bond established with Jesus. And because of Jesus we will get the attention of God and He will be glorified. With this truth we can turn the world upside-down. The early church understood the wonders of prayer. While the believers were praying in the house of John Mark, the angel of God was delivering Peter from prison. Here is the Bible account of the event: *"Peter, therefore was kept in prison: but prayer was made without ceasing of the church unto God for him"* (ACTS 12:5). It is no surprise that Peter witnessed a supernatural deliverance. There is someone who is standing at the right hand of God who is anxiously waiting to take your request and do for you what He has done for others. That link, that bond, that relationship needs to be established through prayer.

4

The Fourth Wonder of Prayer
YOU CAN ENTER GOD'S PRESENCE

The long arm of prayer can reach all the way to heaven. *"Let us lift up our hearts and our hands to God in heaven"* (LAMENTATIONS 3:41 NIV). *"Hear my cry for mercy as I call to you for help, as I lift up my hands toward your Most Holy Place"* (PSALM 28:2 NIV). The expression of lifting our hands up or lifting our hearts up is for the purpose of entering into God's presence. It is the act of our faith that God is alive and that we can have a relationship with Him. The reason why many feel that their prayers go no higher than the ceiling in their room is because they feel that they cannot approach God. God has made a provision that through His Spirit and through the name of Jesus our prayers can enter into His presence. Jesus instructs us to say when we prayer: *"Our Father which art in heaven, Hallowed be thy name"* (MATTHEW 6:9). If we do not physically raise our hands, it is important that we raise our spiritual eyes and see a God who is in heaven that delight in answering our prayers.

The veil in the Temple is a perfect example of the wonder of prayer. For many years the only access a believer in God had was through the intervention of a High Priest. Then the miracle of all miracles happened when Jesus said *"It is finished,"* It had just begun for us. The veil in the Temple was torn from top to bottom, and for the first time a believer could personally go into the Holy of Holies with praise and requests. Prayer ushers you immediately into the presence of God. You don't have to sit in a waiting room, you don't have to make an appointment. There is spiritual reality to prayer. Not only do you have the privilege of touching God, but you can touch another person through prayer. You can span the miles and touch someone who needs help. This is not a spiritual fantasy, this is a spiritual reality. We have all shared experiences of how we were awakened in the middle of the night with a burden to pray for a loved one or friend, then to discover later that the person we prayed for was going through a crisis. I have personally felt the prayers of my congregation praying for their pastor. This is the wonder of prayer that we can touch the needs of others that are miles away.

Bill Wilson, a missionary from South Africa tells an exciting story. He and his brother Willard were going to a new territory in the dangerous Zulu country. They were invited to a village in the bush country to start a new work. They were given a hut to sleep in for the night. A powerful cannibal tribe had heard about the missionaries that had come to their area and they were determined that night to invade the hut and take the lives of the Wilson brothers. It was 3 a.m. when the natives came to the hut, and to their amazement they saw strange creatures that glowed in the night and in their hands were swords. They fled in fear. The next day at a tribal meeting these men came with trembling hearts, and after they heard the story of Jesus they were all converted. Two mission compounds were started as a result of that miracle. About a month later Bill Wilson received a letter from a prayer partner in Brooklyn, N.Y. who stated in the letter that on a certain Sunday afternoon a few people gathered for prayer, feeling the need to pray for Bill Wilson. When they compared the day and time, it was the exact hour of crisis for the missionaries Bill and Willard Wilson. Through prayer, someone in Brooklyn, N.Y. was interceding for someone on the other side of the earth.

Here is the wonder of prayer. We who are weak, finite human beings, saved by the grace of God, are given the privilege of reaching out hands to heaven and touching God's throne. Prayer enables us to reach God's throne with one hand and the needy world with the other hand! Through prayer you can touch someone in California. You can touch someone in Europe. You can touch someone on the outside of your community, and you can do it through prayer in the Spirit. I believe the power of prayer has no limitations. What an honor and privilege it is to come directly into the presence of God. It is an awesome thought that we can come into the presence of a Holy God. We are not worthy of such an honor. It is because of Jesus that this ministry was given to us. It is an open door that is available to us and we can come before God and make our petitions known to him.

Jeremiah caught the Wonder of prayer. *"Call to me and I will answer you and tell you great and unsearchable things you do not know"* (JEREMIAH 33:3 NIV). May we never forget who we are and who He is, and His plan for us is that *"if children, then heirs, heirs of God and joint-heirs with Jesus Christ"* (ROMANS 8:17). Though we don't feel worthy, He honors us by making us part of His kingdom. If you will pray, He will make this ministry important by wrapping His arms around your weak and frail body and make you feel strong and mighty. He will make you feel like a child of the King!

The skeptic has a knock-knock joke about prayer. "Knock, Knock, who's there?" "Nobody!" Jesus has a different version.

"Knock, Knock, who's there? "Our Great God who is mighty and able to do all things is there." Jesus taught us, if you knock, the answer will be opened to you. The skeptic is wrong, if you will just knock on the door of God, that door will open. God is there all the time ready to commune with you.

There was a man who visited a hydroelectric plant. He noticed the water going over the dam at a rapid speed. He asked the tour guide, "What percent of the power from this river do you actually transform into electricity?" The guide said, "We use a small fraction of the power from this river." When I think of the potential that is available to us through prayer, the tragedy is that we are using just a fraction of the resource of power made possible through the presence of God with us. We need to explore this wonder and take advantage of our free access to God to do signs and wonders.

FIVE

The Fifth Wonder of Prayer
YOU HAVE ULTIMATE POWER IN PRAYER

God honors us by providing the freedom to seek his counsel at any time so that we may receive and be empowered by the Holy Spirit to move and perform ministry for Him. The Holy Spirit is in us, and we have a direct access to God to be the recipient instantly of His power. *"For the eyes of the Lord are over the righteous, and His ears are open unto their prayers"* (I PETER 3:12). Again the Bible speaks of this indescribable power of prayer. *". . . pray one for another, that you may be healed. The effectual fervent prayer of a righteous man availeth much. Elijah was a man subject to like passions as we are, and he prayed earnestly that it might not rain: and it rained not on the earth by the space of three years and six months. And he prayed again, and the heaven gave rain and the earth brought forth her fruit"* (JAMES 5:16-18). We are seeing great changes physi-cally, spiritually, financially, because we are using prayer not as a form of some religious tradition, we are using it as a power.

The Christian has a witness of the Holy Spirit which claims, *"I can do all things through Christ which strengtheneth me"* (PHILIPPIANS 4:13). As a result of our prayer relationship with God, the Christian can boldly proclaim. "I can! I will! I shall! I am an overcomer! I have power not only over the devil, but I have power to live a normal balanced consistent life!" In one of our Sunday celebration services at Full Gospel, a women came forward for salvation. Her life had been shattered by drugs and a bad marriage. In a moment through the power of conversion she was a new creation: old things had passed away. At the close of the service I asked her, "Did you break through this morning?" She said "Yes!" Then she said, "I have such a low esteem of myself, I feel selfish to ask God to do something for me." I said to her "That is exactly what the devil wants you to feel, that you are unworthy, that this power is always for someone else. Satan has lied to you all of your life. The opposite is true." With hand raised to God, she found total freedom and deliverance.

Oh Christian! Pull your shoulders back! You have an indescribable power! God wants to make you feel important! He wants to make you feel part of His creation! God will give you a reason to live! He has the power to care for you and to protect you. God does not want you to live beneath His plan for you. God wants you to prosper in all things. He does not want you to live with a poverty mentality. Prosperity is more that riches. Prosperity is being content in whatever condition you are in. You can have an inner power that will stabilize you with things or without things. Money doesn't rule you, you can rule money! The home is not ruling you, you are ruling the home. You have a power over the elements of this world through prayer. You have an indescribable power over the financial world, and over the physical world. It is time you utilize this power for your good and live like a child of God. It is very possible that the reason you are not seeing the abundance of this power can easily be traced to the lack of prayer in your life.

Here Are Six Ways Prayer
Reveals A Unique Power

Prayer-Power To Overthrow And Defeat Satan. Satan is our worst enemy. He is a destroyer. He looks for a way to destroy the work of God and man. It is a wonder to me that at the mention of the name of Jesus we have power over Satan. If God's people would only accept their sacred role as a member of the army of God, if they would call one another to God's priority of prayer, if they would join one another in a militant, Spirit-guided, Spirit-anointed intercession, they could see Satan defeated. Through the power of the blood of Jesus, you can stop Satan and drive him back. If you allow Satan to play in the playground of your heart, then you are lacking prayer. For prayer gives you an authority over the enemy of your soul. James the half brother of Jesus gives us the proper biblical formula to defeat Satan. *"Submit yourselves therefore to God. Resist the devil and he will flee from you"* (JAMES 4:7). You cannot resist the devil unless you first submit yourselves to God. God gives you the power to overcome.

Prayer-Power To Exceed The Physical Laws. If it is God's will and in his divine plan, prayer can exceed the laws of nature. God Himself is a miracle! If God is in your prayer, then a miracle presence is in your prayer. You don't have to seek for miracles, He is a miracle. As you apply God, there is a power that can transcend the laws of nature. Jesus demonstrated this power of nature. We do not have the power in ourselves over nature, but we do have the power to arouse the one that can do it. God can do it. The disciples were fearful in a storm on the sea of Galilee. They thought they were going to perish. In themselves they were helpless, but they awoke Jesus and He said, *"Peace, Be Still."* Yes it was the power of Jesus, but it was

123

also the intercession of the disciples to call upon Jesus for help. I don't have the power to feed the multitudes, but I do have the power to beseech the one that can feed the multitude. His name is Jesus; He has the power to transcend the laws of nature.

When the architect penned our new sanctuary on paper, he could have easily placed it two feet west and missed a solid rock ledge. But he didn't. When the builder, Tom Hamilton, was faced with this rock, he suggested that we alter the plans, and eliminate the two-floor office area that is behind the platform area and make it one floor. By doing this we would bypass the rock. During the evening the Elders and I anointed the rock ledge in the name of Jesus. The next day, the builder affempted one more time to break up the rock, and to our amazement the rock could be broken enough so that a footer could be poured. God is a God of miracles over the laws of nature!

Prayer-Power To Call Upon And Reach God In Heaven. Prayer opens the windows of heaven. The unique ministry of prayer not only brings God down to us, but brings us up to God. It seems that some Christians are powerless. They do not know how to call upon God. We need not make a prior appointment with God, He is never too busy. In prayer, we come to God as His child. There isn't a certain prayer position that is more sacred that gives you greater power. We just need to come to Him in childlike faith for He is eager and willing to see us. Use the power of prayer and come to your God, your Father.

Prayer-Power To Receive Divine Intervention. The Bible teaches us that angels are sent to serve. In today's world people are much more likely to question the existence of angels.

Several people in the Bible have experienced angels as *"ministering spirits sent to serve"* (HEBREWS 1:14). When Jesus was in the Garden of Gethsemane an angel visited him. *"And there appeared an angel unto him from heaven, strengthening him"* (LUKE 22:43) Ask God for protection. Before your children leave for school, cover them with the blood of Jesus and believe God for protection. Whenever you take a trip, ask God for divine protection. At the beginning of a new day, expect the divine guidance and protection that comes to the children of God. This to me is a wonder that in prayer we can feel the witness that He is with us even to the end of the age.

Prayer-Power To Unite And Work With God. God has a plan and He will use the obedient believer even though he or she may not be qualified. I am amazed at this power; that we can work with God. We may lack special skills or training, but we can always work with God through prayer. God can take the weak to confound the wise. He takes those who have feeble lips and uses them to speak. Many times those who can, won't, and those that cannot, will. Do you know why that is so? Because those who do not have the ability recognize through prayer they have a power to cooperate with God. Don't deny yourself the privilege of this power.

Prayer-Power To Be A Blessing To People. We are to follow in Jesus' footsteps and do good wherever we can. *"Jesus went around doing good"* (ACTS 10:38). Like, the blessed Lord, we are to go through life blessing everyone we can. You can bless someone, or you can curse someone. You can lift someone up or you can kick someone down. You can make someone feel wanted or you can reject someone. It is all in how you distribute your power. When you extend your hand, it becomes a power to bless. You might say, "If I had a million dollars I would bless people with it." You may not have a million dollars; but you do have a million dollars of blessing stored in your heart. Prayer will teach you how to give it away freely. Jesus clarifies the way we should bless people. *"And as ye go, preach, saying, The kingdom of heaven is at hand. Heal the sick, cleanse the lepers, raise the dead, cast out devils: freely ye have received, freely give"* (MATTHEW 10:7-8). If you talk to God, he will tell you who to love. If you talk to God He will show you who to forgive. If you talk to God He will tell you who to give to; but if you are not in prayer how can you bless people? What a wonder that God will choose you to bless someone.

6

The Sixth Wonder of Prayer
YOU CAN BE ONE OF GOD'S INTERCESSORS

We are part of a worldwide prayer team. The plan of God prayer is unique because it involves beleivers that united together to share the needs of others. Paul exhorts the Galatian church: *"Bear ye one another's burdens, and so fulfil the law of Christ"* (GALATIANS 6:2). It is a privilege to be a part of a world-wide prayer team. Something is happening through prayer around the world and we are part of it. The secret of growth within the local church is attributed to the intercessors praying for the young christians and the needs within the local church is attributed to the interecessors praying for the ytoung Christians and the needs within the body of Christ. *The Spirit also helpeth our infirmities: for we know not what we should pray as we ought: but the Spirit itself maketh intercession for us with groanings which cannot be uttered"* (ROMANS 8:26). God's Spirit bears witness with our spirit. There is a network, a prayer chain—God working with us with signs and wonders.

There is a tremendous move of God that is occurring around the world and we can be part of God's anointed intercessors. There is only one Holy Spirit and I believe that the same Holy Spirit that is leading me is leading other people to be intercessor. Here is the wonder of this prayer network, that the same Spirit that bears witness with my spirit, will bear the same witness with others others within the body of Christ. The same Spirit that is talking to me is talking to you. When you are in God's prayer network, you will have a burden for people and will intercede for them even when you are not aware of their problems. You will feel an urge to pray for people in various countries or trouble spots around the world. A loved one or a friend may be in the hospital and God will raise you up to be the strength or the encouragement that the person needs.

Within the prayer network you will feel the urgency to pray for the Church, or spiritual leaders, or for the backslidden Christians that are in the valley of decision. Through prayer you will stand between the living and the dead. It will be an honor, privilege and wonder, that the Spirit of God will be within you interceding for others with groanings and words that cannot be uttered by the human language. It is the cry of the Holy Spirit within you. This is praying in the Spirit, a network of prayer that is around the world. This is what God meant when He said: *"Call unto me, and I will answer thee, and show thee great and mighty things, which thou knowest not"* (JEREMIAH 33:3). God desires to use you and me as workers in His Kingdom. Our prayer should be, "O God if you are going to tug at human hearts, tug at my heart and use me as a vessel to reach a needy person that you desire to deliver."

This is a wonder to me because we are weak and have personal needs, but in spite of our shortcomings we can be used of God to pray for other people. It is a distinct privilege to have the Spirit groan within us with words that cannot be uttered. Our desire should be, "O Lord, use me to reach the lives of people and fulfill your Word in me that the prayer of faith shall save the sick." Can you imagine what would happen to our world if Christians neglected to live in the Spirit, refused to be led by the Spirit, and muzzled the mouth of the Spirit? We would literally tie the hands of Almighty God and allow the Devil free and complete access upon the weak, the broken, the hopeless and helpless of our world.

Jesus in the early days of His ministry, went into the synagogue and opened the Book and read, *"The Spirit of the Lord is upon me, because he hath anointed me to preach the gospel to the poor, he hath send me to heal the brokenhearted, to preach deliverance to the captives, and recovering of sight to the blind, to set at liberty them that are bruised, to preach the acceptable year of the Lord"* (LUKE 4:18-19). Jesus was motivated by a divine purpose in His life. He had a job to do, a ministry to develop, a world to reach, and people to help. He knew how the prayer network operated. The Spirit of the Lord was within Him and anointed Him to work the signs and wonders of God upon this earth.

We are witnessing an unusual move of God throughout the world through prayer. In churches and communities God's people are uniting together in spiritual warfare for the souls and lives of men and women. We have seen a clear pattern in the Bible of how God operates. When you are in tune with God's Spirit, He will operate through you. Over and over again we read in the Old Testament that *"the Spirit of the Lord came upon....."* (JUDGES 3:10, JUDGES 6:34, JUDGES 14:6, 1 SAMUEL 16:3, ISAIAH 11:2, EZEKIEL 11:5). But now in the new covenant we read that *"the Spirit of God is within us"* (MATTHEW 3:16, ROMANS 8:9, 1 CORINTHIANS 3:16). Why! Because He wants us to be part of His prayer network and desires to move throughout the earth.

What a wonder! We can make an endowment and an

129

investment in the lives of people. In our church we distribute a prayer sheet that sometimes has over five pages of personal request of needs that people have turned in. I am constantly aware that when you pray you are making an investment of your time and energy with every individual person. *"Pray one for another that you may be healed"* (JAMES 5:16). In prayer there is unity, there is love, there is forgiveness, there is strength, there is power, and there is encouragement. Nothing can unite us greater than the ministry of prayer. When each of us pray we are bound together by one Spirit, and being part of that network we will be one in the Spirit. When one rejoices we all rejoice, when one cries we all cry, when one has a burden we all have a burden. Prayer works out that miracle among us. We can be touched by the feelings of the infirmities of others. Only prayer can do this. When we are linked together something beautiful will happen.

How do we Become One of God's Intercessors?

First, when you are spirit-filled you will be willing and ready to do God's will. Not every Christian has a spirit-filled and spirit-controlled life. The believer must make a "total surrender" and that surrender is to the Holy Spirit. To be spirit-led, spirit-controlled, spirit-filled means "you are available." God can knock on the door of your heart at anytime and you will be ready *"in season, or out of season"* to do the will of God (2 TIMOTHY 4:2). You and I can be in touch in God, because we are filled with the Spirit.

Second, seek ways to improve your prayer life. It just doesn't happen overnight. You must work at prayer. The more you experience the life of prayer, the more God will be able to demonstrate his power to you. You will learn how to flow with God. Commit yourself to spend premium time with God. Take prayer seriously.

Third, learn how to communicate with God in a meaningful way. It is not learning to pray by saying religious expressions or religious clichés, but rather learning how to talk to God from the sincerity of your heart. We need to talk to God with expressions that we are comfortable with, and in a language that we understand. In prayer it is not that we are trying to impress God with our words, but rather making contact with God through natural conversation. God is your Father, and He delights for you to live in His presence.

Fourth, be attentive to God when he speaks to you so that you will respond immediately. We must learn to listen to God. We need to know when He is speaking. We can't be so busy and easily distracted. Prayer is not communion if you do all the talking. Listen to the still small voice of God. The basis for obedience is being able to hear from God and being ready to obey Him. We easily hear from men, but we need to hear from God. This prayer network establishes listening ears so that we are sensitive to the heart-cry of God and the cries of a needy world.

Fifth, seek the help of God to keep you vigilant. There are far too many sleeping Christians. Keep a prayer list. Be sensitive to any new prayer assignment God may point out to you. *"Stir up the gift of God, which is in thee"* (2 TIMOTHY 1:6). Prayer is a way of life. The Bible says,*"men ought always to pray and not faint"* (LUKE 18:1). This tells me that there is an activity that we must keep alive. Your prayer can be an eternal investment. Heaven's records are adequately kept. God is not so unjust as to forget your work and your love you have shown Him, by helping others. *"For God is not unrighteousness to forget your work and labor of love, which ye have shown toward His name, in that ye have ministered to the saints, and do minister"* (HEBREWS 6:10). God's word is clear in exhorting us that your ministry of intercession for the salvation of loved ones and friends, and for the building of Christ's Church, and for a spirit of revival among God's people, and for the evangelization of the world, all this and more, *will have eternal rewards.* God does not forget! Someday in eternity He will remember your ministry of intercession in prayer and reward you openly. The one quality that makes prayer a wonder is that every investment you make in prayer will remain with you forever. It is the greatest investment you can make in your life.

Prayer is an intensely personal investment of love. Prayer is the only expression of love which cannot be stopped or rejected. Though the power of prayer you can even surround with love the lives of those who reject you. This is why God's Word in Matthew 5:44 emphasizes to pray for them that persecute you and do all manner of evil against you. Why is praying so important? Because, they may not accept your love, but they cannot destroy it. It is all stored up in the treasure house of heaven and in the memory of God.

Here is a wonder of God's intercessory network. It can build up an endowment of blessing for you! Your prayers have no limitation, they can go around the world. Don't treat your prayer responsibility as legalistic bondage and feel you can never take time to relax, chat with others, enjoy God's nature, or have time with the family or friends. Prayer should never distract you from the natural responsibilities of life. Prayer should never be an excuse to relieve you from the necessities of life. Prayer is a spiritual encounter with God. It is done in the spirit of a person, even when they are involved in physical things. Think of all the countless minutes that are wasted each day by the average person, which could be invested in prayer. Imagine, you can spend five minutes in prayer for someone, and because of your prayers they will be in eternity forever with God.

What a tremendous investment you can make in someone's life! I believe in eternity someone will come to you and hug your neck and say, "If it wasn't for your prayer ministry, I would not be here. Thank you! It was your love, your witness, your life, your testimony, that helped me!" That is an investment. If you love someone, invest prayer in that person. If you love your church and pastor, then pray sacrificially for them. Someday Christians will receive from God a crown for their faithfulness, as recorded in the book of Revelation, and they will *"cast their crowns before the throne, saying, Thou art worthy, O Lord, to receive glory and honor and power"* (REVELATION 4:10-11). What a glorious moment that will be when God's intercessors from around the world will be rewarded. Expect rich dividends! Prayer has made you rich!

SEVEN

The Seventh Wonder of Prayer
THE ALL-INCLUSIVE CONDITION OF PRAYER

In all of God's dealings with us, His promises and His conditions are inseparable. If we fulfill the conditions, God will fulfill the promise. This is a wonder of prayer. Draw near to God and He will draw near to you. Jesus explains this wonder. *"If ye abide in me, and my words abide in you, ask whatsoever ye will, and it shall be done unto you"* (JOHN 15:7). The promises of God are conditional. Prayer is not like an automatic dispensing machine into which we deposit our requests and receive what we desire. Prayer can be hindered. Peter warned husbands and wives to maintain good relationships, *"in order that your prayers be not hindered"* (1 PETER 3:7). If the Bible says that a relationship between a husband and wife can hinder prayer, then the relationship between you and the church can hinder prayer. The relationship between you and your brother can hinder prayer. The relationship between you and your friends can hinder prayer. What God is saying is that there is a condition in prayer. If you want to see answers to prayer, let your relationship with others be pure, right and wholesome.

Here is the basis for unhindered prayer. Unhindered prayer is the result of a sound relationship between us and the Lord, us and the church, us and the world. Jesus put this in a positive way. *"If you abide in me and my word abide in you, you shall ask what you will and it shall be done unto you"* (JOHN 15:7). The secret to unhindered prayer is to abide in Christ. As the union of the branch with the vine produces growth, a never-ceasing growth in spite of the negative elements, so it is with our prayer union. If we meet the conditions of abiding in Him, in His love and compassion for others, then our prayer life brings growth. We will see results. It is the law of the harvest. As the branch abides in the vine, it will bring forth fruit, much fruit. If you want to see results in your prayer ministry, *"abide."* Abide in Christ, abide in the faith, abide in the love of God, abide in fellowship with your brothers and sisters. We are not spending enough time meeting God's conditions.

Andrew Murray the great teacher of prayer said that prayer consists of two parts, human and divine. The human side is the asking; the divine side is the giving. The all-inclusive condition of prayer is that when we do our part to maintain a union with Christ in prayer, God will fulfill His promises. This is a wonder to me, that you and I can either tie the hands of God or release the hands of God. We tie His wonder-working hands by our disobedience, being out of harmony with God and men, and by living contrary to His word. But thank God we can loose His hands to do great signs and wonders when we have met His conditions of a life of obedience to His will. With a clean heart we can come boldly to the throne of grace in prayer.

Prayer is the breath of the soul. Look at how air is received into our body, and envelops us. The air seeks to enter our body, but we must have lungs to receive it. Just think, all of the potential and power of air wasted if the human body did not have the conditions to receive this air. It is well known that it is more difficult to hold one's breath than it is to breathe. As air enters into our lungs and performs its life-giving functions, so it is with prayer. If you stop prayer, you will stop a major function of allowing air to flow for the breath of the inner man.

To pray is nothing more than to involve Jesus in the needs of our life. To pray is to give Jesus permission to employ His powers in the alleviation of distress. To pray is nothing more involved than to open the door and give Jesus access into the affairs of our lives. To pray is nothing more involved than to stand in the sunshine of His grace, to expose our distress to His healing rays, and then patiently wait on God. Here is the wonder of prayer. To pray is to give Jesus the permission to employ His power in the alleviation of our distress.

PART FIVE

THE WONDER
OF THE HOLY SPIRIT

John 16:7-15

Few Christians seem to realize how vital and important the Holy Spirit is within their lives. They believe in the Holy Spirit as a distinct experience but fail to apply Him to their own personal lives. For this reason there is a vacuum and an emptiness in the hearts of many Christians. The coming of the Holy Spirit is an absolute necessity in carrying out the program of Christ's kingdom on earth. All that we have known, witnessed and experienced in the Christian faith could never be made possible if it were not for the Holy Spirit. Jesus declared that it was expedient or profitable for them that He should go away, because His departure meant the sending of the Holy Spirit to the believers. He stated that, " *It is expedient for you that I go away. If I go not away the Comforter will not come unto you*" (JOHN 16:7). Christ was leaving, but the Holy Spirit was coming. The words of Jesus become words of life to us.

Jesus reveals the purpose of the Holy Spirit. *"Howbeit when he, the Spirit of truth, is come, he will guide you into all truth: for he shall not speak of himself; but whatsoever he shall hear, that shall he speak: and he will show you things to come"* (JOHN 16:13). I am convinced that I can radiate the desire of God through the Holy

Spirit. This is profound! God says to the Holy Spirit, " I want you to share this truth with Al Alaimo" The work of the Holy Spirit is now to reveal the wisdom and knowledge of God that is for my best interest. I must be obedient and submissive to the leading and guidance of the Holy Spirit in my life. This is what the Bible means when it says, *"For as many as are led by the Spirit of God, they are the sons of God"* (ROMANS 8:14). Our prayer should be " Holy Spirit lead me, teach me and guide me into all truths."

It would be well to notice that Jesus felt that the Holy Spirit was so important for the Church that He declared, " *If I depart, I will send him unto you"* (JOHN 16:7). This is a direct promise from Jesus Himself. He promises that there will be someone that will constantly be by your side during troubled times, for He shall physically dwell within you. " *When the enemy shall come in like a flood, the Spirit of the LORD shall lift up a standard against him"* (ISAIAH 59:19).

What if the biblical record of Acts chapters one and two had been dropped from the Bible and the book of Acts would have begun with chapter three? We would be compelled to believe that some event of unparalleled interest and importance had occurred to account for the astounding acceleration of power in the life of Peter and the apostles. You would always wonder where and how did they get this unusual power to live a victorious life? What produced this unusual happening?

When you read the first and second chapters of the book of Acts you see the advent of the Holy Spirit and the indwelling experience of the life of the Spirit within the hearts of the believers. In the lives of the Christians there was an escalation of power and victory and excitement. Prior to the book of Acts the disciples had only limited or delegated leadership or authority. During the last days of Christ's ministry on earth you see them disorientated and confused about the direction of the Kingdom of God. When Peter indicated that he was going back to his occupation of fishing, the disciples agreed to go with him, (JOHN 21:3). They were completely visionless on the day of the ascension of our Lord.

What happened to these followers of Jesus? You cannot have an effect without a cause, without an adequate cause. There was a reason why the people said, *"These that have turned the world upside down are come hither also "* (ACTS 17:6). If one day you see someone lifeless and then the next day you see them filled with excitement, there must be an explanation why the change happened in their lives. If you see 50 train cars standing quietly on the rail and in less than five minutes see them moving rapidly down the track, you will conclude the existence of a power. The only explanation I have in reference to the change that occurred in the lives of the first century Christians is the Holy Spirit. As we observe the Church of the twentieth century we see a phenomenal growth among Spirit-filled believers. It is reported that from 1896 to 1996 over 500 million believers have received the gift of the Holy Spirit. The credit and honor for our growth as a Pentecostal Movement is the Blessed Holy Spirit.

When you read in the Bible about the life of the disciples, in Luke 24, in contrast to the life of the disciples of Acts 3 you are compelled to demand an explanation. Let the mind of a thinking person find an adequate explanation other than the advent of the Holy Spirit for the unusual transformation in the lives of people.

In the book of Luke the disciples were slow to understand the Holy Scripture. In the early chapters of Acts we are amazed at the many scriptures they are quoting. Something had to

141

happen to them to produce the unerring accuracy so that they would apply the scriptures in their spiritual warfare. It is the manifestation of the Holy Spirit that will make the Word of God come alive to your heart. You will hear it, absorb it, and share it with other people. You will know how to apply Scripture to the everyday problems of life.

In the book of Luke, the disciples trembled as they fled from the Master's accusers, but view Peter after the Day of Pentecost in the book of Acts as he stands firm as a rock. He stands in the street of Jerusalem before 3,000 people and boldly charges them with the murder of the Messiah, Jesus Christ. It took courage to be a witness for His Lord, and the credit should be given to the new found gift that was within him. Here is the wonder of the Holy Spirit. One day Peter was weak, struggling to be a Christian and the next day he was ready to defy death through the power of the Spirit. It took courage to be a witness.

In the book of Luke, the disciples had bitter sorrow, half-hearted enthusiasm, little power with God and man. In Acts we read: *"And the multitude of them that believed were of one heart and of one soul. And with great power gave the apostles witness of the resurrection of the Lord Jesus: and great grace was upon them all"* (ACTS 4:32-33). The explanation for the changes in the lives of these Christians was the Holy Spirit.

I have observed during my pastoral ministry a misunderstanding in regard to the Holy Spirit by well meaning Christians. It appears that there will always be theological controversy on the subject of the work and operation of the ministry of the Holy Spirit within the Church. Even though there is controversy concerning the works of the Holy Spirit in the Church today, there should be fundamental truth accepted by all Christians. Why believe in the third person of the Holy Trinity?

We must go back to the very beginning of the miracle of creation to find the answer. *"And God said, Let us make man in our image, after our likeness"* (GENESIS 1: 26). This is one of the most important scriptures in the Bible in order to understand the Holy Spirit. What is the image and likeness of God? The image and likeness of God is not flesh and blood. God is a Spirit. The image and likeness of God is the nature, the personality , the wisdom and the Spirit of God. We were created in the image and likeness of God so that we can have a relationship with Him. *"God is a Spirit: and they that worship him must worship him in spirit and in truth"* (JOHN 4:24). God's image and likeness in the human nature today has been corrupted because of sin. The wonder of the Holy Spirit is that He has come to bring restoration and fellowship with God to our lives .

To add revelation to this truth, Paul reveals in Romans 8 why we need the Holy Spirit. It is because we were born in sin that the nature of God within us is dead. The psalmist David was aware of the weakness of the human nature when he said: *"Behold, I was shapen in iniquity, and in sin did my mother conceive me"* (PSALMS 51:5). The gift of the Holy Spirit provides a great miracle within us. *"There is therefore now no condemnation to them which are in Christ Jesus, who walk not after the flesh, but after the Spirit. For the law of the Spirit of life in Christ Jesus hath made me free from the law of sin and death"* (ROMANS 8:1-2).

Since the Holy Spirit is resident within us, we can be led by God's Spirit instead of the human spirit. The purpose of the Holy Spirit is to guarantee that we live the abundant life. He promises us that we will no longer walk after the flesh but, *" after the Spirit"* (ROMANS 8:4). He assures us in verse 9 that the Spirit of God will *"dwell in us."* He tells us we will never be lost, because we will be, *"led by the Spirit"* (verse 14). We will never feel lonely because (verse 16) the Holy Spirit will *"bear*

143

witness with our spirit." In the times of sickness, verse 26 says, the Holy Spirit will help, *"helpeth our infirmities."*

The human body needs some help. Our emotions are shattered. Our confidence is crumbling. The temptations around us are overwhelming. How do we live in a negative world? We live with a positive understanding that the Holy Spirit is alive within us. Adam and Eve were created to have fellowship with and communicate with God. Through the power of the Holy Spirit we can again renew that relationship with God. We can now talk to God, feel Him and know that He is alive.

1

The First Wonder of the Holy Spirit
HE GIVES INSIGHT INTO SCRIPTURE

There will come a flash of new meaning to the words of the prophets, the apostles and the Bible stories. We cannot interpret the Scriptures with our own understanding. If we do, then we would be reading the Bible as we read a novel. The enlightenment that comes to us is from the Holy Spirit. When you read the Psalms, the words are not just poetry they are God's spiritual therapy to an aching heart. When you read the events in the book of Exodus it is not just millions of Jews leaving Egypt. The Bible is a living book filled with practical and personal application to our Christian experience. When the Holy Spirit comes He shall illuminate the dark chambers of your soul and spirit.

The same Holy Spirit that gave the Word will give you insight to the Word. *"For the prophecy came not in old time by the will of man: but holy men of God spake as they were moved by the Holy Ghost"* (2 PETER 1:21). He who wrote the Book will give you interpretation to its words and meaning. The same Spirit that gave the Word will reveal the Word. Your heart will be open to understand the Scriptures and your eyes will see with divine insight. The old chapters will assume new interest. The trivial incidents will take on new importance and historical events will have deeper meaning. It is amazing to me that as many times as we have read a familiar scripture there is always a new truth that is revealed to us from it. For example, you may have read Psalm 23 hundreds of times, but in your hour of a particular need that portion of Scripture will reveal a special message for you that is needed during your present problem. The Bible can be just words, but when the Holy Spirit comes within you they are more than words. They have new meaning and new understanding because of the insight of the Holy Spirit.

When the Holy Spirit comes to us, we find the miracle of transfiguration repeated and the Word of God lives within us as a divine power. Transfiguration means that something physical is changed into something divine. Peter, James and John were awe struck when they were on top of the mountain and saw the transfiguration of Jesus with Moses and Elijah. The Bible says, *"While He yet spake, behold, a bright cloud overshadowed them: and behold a voice out of the cloud, which said, This is my beloved Son, in whom I am well pleased; hear ye him"* (MATTHEW 17:5). Peter was so impressed with what he saw that he wanted to build three tabernacles for Jesus, Moses and Elijah. He missed the whole point of the transfiguration experience. Jesus wanted to teach them that it is possible through the insight to Scripture that we can see divine revelations. The Word of God is designed by Almighty God to be transfigured. To many people the Bible is just words; but to a believer these are words of life and hope. The Bible reveals that the Holy Spirit did not come to teach a new message but to bring to the mind of a Christian the message that Christ taught and to give them the spiritual understanding of it. *"And the Word was made flesh, and dwelt among us, (and we beheld his glory, the glory as of the only begotten of the Father,) full of grace and truth"* (JOHN 1:14). We need to behold the message from our Father. Allow the Holy Spirit to transfigure and glorify God's Word.

The Holy Spirit is the author of the Word of God and for this reason a Christian has a greater understanding of its content. *All scripture is given by inspiration of God, and is profitable for doctrine, for reproof, for correction, for instruction in righteousness: That the man of God may be perfect, thoroughly furnished unto all good works."* (2 TIMOTHY 3:16-17). When one is intimately acquainted with the author of a book, then he enjoys the book better. Many people have not enjoyed the Bible because they have not become acquainted with the author. The author of the Bible is God the Holy Spirit. *"Howbeit when he, the Spirit of truth, is come, he will guide you into all truth: for he shall not speak of himself; but whatsoever he shall hear, that shall he speak: and he will show you things to come"* (JOHN 16:13).

Many things in the Word of God are spiritually discerned. This is how Paul describes the operation of the Holy Spirit in reference to the Word of God. *" But as it is written, Eye hath not seen, nor ear heard, neither have entered into the heart of man, the things which God hath prepared for them that love him. But God hath revealed them unto us by his Spirit: for the Spirit searcheth all things, yea, the deep things of God. For what man knoweth the things of a man, save the spirit of man which is in him? Even so the things of God knoweth no man, but the Spirit of God. Now we have received, not the spirit of the world, but the spirit which is of God; that we might know the things that are freely given to us of God. Which things also we speak, not in the words which man's wisdom teacheth, but which the Holy Spirit teacheth; comparing spiritual things with spiritual. But the natural man receiveth not the things of the Spirit of God: for they are foolishness unto him: neither can he know them, because they are spiritually discerned"* (1 CORINTHIANS 2:9-14). Here is the wonder of the Holy Spirit. God has given us a Book that we cannot understand with the natural mind, but when the Holy Spirit illuminates the words, it comes alive with meaning and understanding.

The great need of our day is insight to Scripture and we can have it through the wonder of the Holy Spirit. **The Word of God** can become a lamp into our feet and a light to our pathway (PSALM 119:105). **The Word of God** is pure and a shield to them that put their trust in Him (PROVERBS 30:5). **The Word of God** is of none effect through your tradition (MARK 7:13). **The Word of God** is the seed (LUKE 8:11). **The Word of God** grew and multiplied (ACTS 12:23). **The Word of God** prevailed (ACTS 19:20). **The Word of God** is the sword of the Spirit (EPHESIANS 6:17). **The Word of God** is not bound (2 TIMOTHY 2:9). **The Word of God** is quick and powerful (HEBREWS 4:12). **The Word of God** abides in you (1 JOHN 2:14). The last triumph in the book of Revelation is the Word of God. *"And I saw heaven opened, and behold a white horse; and he that sat upon him was called Faithful and True, and in righteousness he doth judge and make war. His eyes were as a flame of fire, and on his head were many crowns; and he had a name written, that no man knew, but he himself. And he was clothed with a vesture dipped in blood: and his name is called **The Word of God** "* (REVELATION 19:11-13). Jesus is the revealed Word.

2

TWO

The Second Wonder
of the Holy Spirit
HE GIVES GREAT BOLDNESS
OF UTTERANCE

When we examine the life of Peter we observe that during the false trial of Jesus Christ He trembled like an aspen leaf before the questions of a maid. Later we see him standing like a lion, preaching with great boldness the message of God. We are quick to note that there was a spiritual encounter in his life. There was no hesitancy in speech, no cringing in demeanor, and no fearful actions before the people. Peter spoke with the authority of the Holy Spirit. Boldness is not harshness, nor the creating of a hostile atmosphere. Boldness is to be led by the Spirit and to say what He wants said to a needy and spiritually hungry world. The wonder of the Holy Spirit is to see Christians who do not have the physical ability to speak with authority and boldness, do so after they receive this gift. The words they say will not be their words, but the words of God spoken through their lips.

The primary purpose of the advent of the Holy Spirit is not just to enable people to speak with new tongues as Paul testified to in his life: *"I thank my God, I speak with tongues more than ye all:"* (1 CORINTHIANS 14:18). Paul taught that we will receive a new tongue of boldness so that we can become a witness. *"God also bearing them witness, both with signs and wonders, and with divers miracles, and gifts of the Holy Ghost, according to his own will"* (HEBREWS 2:4).The prayer language is only a witness or evidence to us that the Comforter has come to abide in the life of a believer. Boldness of utterance is to be a witness to an unbelieving world where you have never been a witness, and to declare God's truth where you never thought you would declare God's truth. Witnessing for Jesus and boldness of utterance definitely accompany the Spirit-filled life.

There is an irresistible passion to tell others about the Savior that grips the life of a person who has the Holy Spirit. Jesus describes the sequence of events in the life of a Spirit-filled believer. *"But ye shall receive power, after that the Holy Ghost is come upon you: and ye shall be witnesses unto me both in Jerusalem, and in all Judaea, and in Samaria, and unto the uttermost part of the earth"* (ACTS 1:8). In the very beginning of the advent of the Holy Spirit, God's design was that the followers of Christ would have a boldness to communicate the Gospel to all people. Have you reached the stage of holy boldness so that your words with love and kindness will rebuke the actions of a wicked person? Have you experienced giving the right words, at the right time to a careless individual? Have you felt inadequate in yourself but bold in the Spirit to share the good news of the Gospel to someone who was lost and hopeless?

How do we explain this boldness? Something happened to them in the Upper Room. *"Now when they saw the boldness of Peter and John, and perceived that they were unlearned and ignorant men, they marveled; and they took knowledge of them, that they had been with Jesus"* (ACTS 4:13). This is the way Peter described the pentecostal story to thousands of people that assembled in the street of Jerusalem. *"And when they had prayed, the place was shaken where they were assembled together; and they were all filled with the Holy Spirit, and they spake the word of God with boldness"* (ACTS 4:31).The boldness that Peter is referring to is not a learned gift; it is an empowered gift that comes from the Spirit. You can only be an effective witness of something you have experienced for yourself. The Holy Spirit will cause things to come alive within you and He will give you the boldness of utterance to share the message of God with others. We are a witness of Christ. The best teacher you can have to be an effective witness is the Holy Spirit.

The characteristic of a Spirit-filled Christian is that when someone has a need there is an immediate response. You can say to someone, "You must be born again." Your words do not change them. The boldness of utterance is when the Holy Spirit leads you to speak with wisdom and understanding, the hopeless will find themselves compelled to listen. The ungodly are pricked in their heart when you speak. It is not only the words you say, but also the power by which it is received. It is the power of the Holy Spirit that makes the difference. *"This is the word of the LORD unto Zerubbabel, saying, Not by might, nor by power, but by my spirit, saith the LORD of hosts"* (ZECHARIAH 4:6). Here is the wonder of the Holy Spirit, you will have "tongues *like as of fire"* (ACTS 2:3). Tongues like as of fire is not only having powerful truth from God, but a powerful Spirit behind your simple words.

The Third Wonder
of the Holy Spirit
HE PRODUCES JOY AND
FERVENT LOVE

The disciples only ten days before had lost the presence of their dearest friend Jesus Christ. They saw Him ascend up into heaven and were told to go to the Upper Room to wait for the Promised Gift of the Holy Spirit. I have discovered that the disappointments in life produce a vacuum. Many times when you lose a special friend it will rob you of the love for life and the joy for living. I can imagine, 120 confused, sad, empty believers waiting for something to happen that they could not explain. Jesus had a purpose for His followers. He knew that the joy of God that surpasses all understanding would replace the joyless life they were experiencing. They must have said to each other, "What are we doing here in the Upper Room?" "What are we waiting for?" Possibly a believer said, "I am not leaving until I receive the full knowledge of the kingdom." They wanted the kingdom to come. *"For the kingdom of God is not meat and drink; but righteousness, and peace, and joy in the Holy Spirit"* (ROMANS 14:17). When the Holy Spirit came upon them, they were filled with ecstasy of love, peace, joy and hope. God did not send the Holy Spirit so that His

Church would have a yearly celebrated holiday called, " Pentecost Sunday." Pentecost is God's mark upon a human life to bring true joy and love.

This experience became a wonder to the bystanders in Jerusalem. Peter tried to explain the demonstration of their love and joy. He said: *"For these are not drunken, as ye suppose, seeing it is but the third hour of the day. But this is that which was spoken by the prophet Joel; And it shall come to pass in the last days, saith God, I will pour out of my Spirit upon all flesh"* (ACTS 2:15-17). When they were given such an explanation, there must have been an outward demonstration of the excitement of joy. Everybody was having a great time of rejoicing. Here is the wonder of the Holy Spirit, when He comes He takes you out of yourself and He fills the void that is within your life You do not need drugs or liquor to bring you joy; the Christian has a better solution, it is the Holy Spirit. To obtain joy in life, the unbeliever has to buy and seek for it and many people destroy themselves to try to get it. The Christian has received God's joy without price, because it is a free gift through Jesus Christ.

This is how Paul describes the gift of the Holy Spirit. *"And be not drunk with wine, wherein is excess; but be filled with the Spirit; Speaking to yourselves in psalms and hymns and spiritual songs, singing and making melody in your heart to the Lord; Giving thanks always for all things unto God"* (EPHESIANS 5:18-20). What does it mean to be filled with the Spirit? One of the things that it means, is that we will speak to ourselves with psalms and hymns. It is when the spirit life is more predominant than the self life that we experience the abundant life. You may be filled with gloom and sadness, but you can be lifted up in the Spirit and have a melody in your heart. Joy is the outflow of the Spirit-filled experience. There is human joy that is produced by the physical, but there is a spiritual joy that is produced by the

Holy Spirit. It is not something that happens from you, but it happens from the One that is in you. A physical joy is temporary, but God's joy is permanent.

The prophet Nehemiah said unto the people, *"For the joy of the LORD is your strength"* (NEHEMIAH 8:10). The prophet Isaiah says, *"Behold, God is my salvation; I will trust, and not be afraid: for the LORD JEHOVAH is my strength and my song; he also is become my salvation. Therefore with joy shall ye draw water out of the wells of salvation"* (ISAIAH 12:2-3). Joy is the bucket that dips down into God's reservoir of spiritual blessing. There is no way you can draw water out of a well without a bucket. The Holy Spirit creates the container of joy and God fills it with His presence.

God wants us to rejoice! That is why He has given us the Holy Spirit. Paul was in prison and he wrote words of encouragement to the Church: *"Rejoice in the Lord always: and again I say, Rejoice"* (PHILIPPIANS 4:4). One of the characteristics of the Holy Spirit is the spirit of joy. We sing a praise chorus from Psalms 100:4 that explains our new life. I will enter His gates with thanksgiving and praise. I will rejoice for He has made me glad.

The wonder of the Holy Spirit is that He gives us the baptism of joy and a fervent love for lost souls. An inner power promotes an outward action. God's kingdom demands that something must first happen on the inside before something can be done on the outside.

FOUR

The Fourth Wonder of the Holy Spirit
HIS PERSONALITY WILL BE WITHIN US

The Holy Spirit is not an " it;" He is a person! The Holy Spirit is not a mere force or just an influence. He possesses all the attributes of personality, intellect, emotions and will. Here is the wonder that He will dwell within us. Several years ago Mary and I were in New York City during the Christmas season walking the streets and enjoying the colorful lights and decorations. There were a few women walking towards us. I knew that they were Spirit-filled. As they got closer, a boldness came on me and I said, " Are you women Spirit-filled?" They said, " Praise God, yes we are! How did you know that among all of these thousands of people walking the streets, that we were Spirit-filled?" I said, " There is something about Spirit-filled people; they have a personality." On the corner of Times Square and Broadway we began to exalt the name and life of Jesus Christ. What makes Spirit-filled people different? It is their attitude, the way they look at things, the life of the third person of the Godhead within them. Here is the wonder of the Holy Spirit, He radiates within us. There is something within our eyes and spirit that is alive. As a Christian it is impossible to have the Holy Spirit within us and be dead and lifeless. He desires to live through us, so that our world will know that God is alive . " *The Spirit itself beareth witness with our spirit, that we are the children of God*" (ROMANS 8:16).

The Holy Spirit Is A Person

The Holy Spirit has intelligence. Jesus promised to send One whom He called the Comforter. "*But the Comforter, which is the Holy Spirit, whom the Father will send in my name, he shall teach you all things, and bring all things to your remembrance, whatsoever I have said unto you*" (JOHN 14:26). Only a being with intelligence could teach and bring things to our remembrance. What is a comforter? He is One who is by my side at all times guiding me and directing me in the ways that I should go. There is a simple but powerful illustration given to us by Solomon one of the wisest men that ever lived. He said: " *Train up a child in the way he should go: and when he is old, he will not depart from it*" (PROVERBS 22:6). It takes an intelligent person to train another person to be intelligent.

The Holy Spirit has personality. If the Holy Spirit was only a mere force or some emotional experience, then we would be limited to know about the knowledge of God and the life we should live. Paul indicated that He, the Holy Spirit is capable of expressing prayers on our behalf. " *Likewise the Spirit also helpeth our infirmities: for we know not what we should pray for as we ought: but the Spirit itself maketh intercession for us with groanings which cannot be uttered. And he that searcheth the hearts knoweth what is the mind of the Spirit, because he maketh intercession for the saints according to the will of God*" (ROMANS 8:26-27). This speaks of His ability to make intercession for us. If He makes intercession for you, then He knows how you feel, and if He knows how you feel then He is a personality. The personality inside us is not a dead form but a living life. Ask the Holy Spirit to teach you, lead you and guide you into the will of God. The personality of the Holy Spirit does not make us depressed; on the contrary, He is an encourager to us. He is not an "it." He is a person. He knows how to relate to us as a person. He gives us assurance of salvation and makes intercession for those who are believers. We can have a testimony that it is not the things we do, but it is the Spirit who works through us. His life radiates through us. " *Let your light so shine before men, that they may see your good works, and glorify your Father which is in heaven"* (MATTHEW 5:16). This is the wonder of the Holy Spirit.

The Holy Spirit has emotions and will. Paul taught that the Holy Spirit has feeling. He can be grieved. " *And grieve not the holy Spirit of God, whereby ye are sealed unto the day of redemption"* (EPHESIANS 4:30). He can be irritated and vexed. When Isaiah relates the story of Israel leaving Egypt he points out why they wandered in the wilderness for forty years. It was not something that God did, but what they did against the Holy Spirit. " *But they rebelled, and vexed his Holy Spirit: therefore He was turned to be their enemy, and He fought against them"* (ISAIAH 63:10). I want the person of the Holy Spirit on my side. The Bible gives us a warning, *"Quench not the Spirit."* (1 THESSALONIANS 5:19). We must be very careful that we do not interfere with the will of the Spirit. Many people are negative by nature and find themselves criticizing others when they really do not want to display such actions. The Holy Spirit will make the difference in their life. He will bring a positive attitude because of His God nature that lives within us. His personality is very positive, very assuring, very solid, and very firm. Who would not want that kind of personality within them? This is a day when people spend billions of dollars for counseling and therapeutic treatment. They need to know that the Holy Spirit will lead them out of the wilderness, and all they have to do is to receive Him.

Just recently I heard someone who was going through a severe trial say, " I do not know what I would do if I did not have the Holy Spirit within me?" We would make wrong decisions without His guidance. It is when your back is against the wall, or you are alone in a hospital or facing impossible circumstances, that you need to feel the presence of the Holy Spirit. It is in these difficult situations that you need to commune with God in the Spirit with your prayer language that the Holy Spirit has given to you. If you have not shared in this experience, you have robbed yourself of the wonder of the Holy Spirit. There is no greater compliment given to you than to have someone say that you radiate with the personality of the Holy Spirit.

5
FIVE

The Fifth Wonder of the Holy Spirit
HIS DEITY HAS BEEN REVEALED

We should always relate to the Holy Spirit as God. He is God, the eternal member of the Holy Trinity. How do we know this? The 5th chapter of Acts records the sin of Ananias and Sapphira, who claimed that they were giving to the apostles the full price of the land that they sold, although they kept some of the money from the sale for themselves. The choice was with Ananias and Sapphira on how much they should give to the Lord. Their sin was that they covered up a lie, and not that they kept back the price for themselves. When they gave their offering in response to the need in the church they said it was the full price and this was not the truth. Here is a wonder, that God's truth would be revealed through the Holy Spirit. *"But Peter said, Ananias, why hath Satan filled thine heart to lie to the Holy Spirit, and to keep back part of the price of the land? Whiles it remained, was it not thine own? and after it was sold, was it not in thine own power? why hast thou conceived this thing in thine heart?"* Then Peter said: *"thou hast not lied unto men, but unto God"* (ACTS 5:3-4). This clearly indicates that the Holy Spirit is God. They did not lie to man,

161

but to God. They were not honest with the Holy Spirit. The conclusion of this story is sad.

"And Ananias hearing these words fell down, and gave up the ghost: and great fear came on all them that heard these things. And the young men arose, wound him up, and carried him out, and buried him. And it was about the space of three hours after, when his wife, not knowing what was done, came in. And Peter answered unto her, Tell me whether ye sold the land for so much? And she said, Yea, for so much. Then Peter said unto her, How is it that ye have agreed together to tempt the Spirit of the Lord? behold, the feet of them which have buried thy husband are at the door, and shall carry thee out" (ACTS 5:5-9). God's purpose for this story to be recorded in His Word is to warn us that we should not say something to the Holy Spirit that is not true. He is God. Therefore we must be very careful how we relate to the Holy Spirit in our lives. I am seriously concerned about the casual attitude that some Christians have in relationship to the gift of the Holy Spirit and spiritual gifts. They should not treat the Holy Spirit and their prayer language with disrespect. We should not bring Pentecost down to human level, but it should raise us up to a high spiritual level.

You cannot live an immoral and sinful life and then expect Deity to dwell within you. The Bible says: *" Know ye not that ye are the temple of God, and that the Spirit of God dwelleth in you? If any man defile the temple of God, him shall God destroy; for the temple of God is holy, which temple ye are"* (1 CORINTHIANS 3:16-17). God dwelling in you, is a sacred experience! When you make room in your body to be the temple of the Holy Spirit, you are making room for God. There is an awesome responsibility when you are involved with God. The Holy Spirit is God and His dwelling place is in the life of the believer. What a wonder that God would condescend to dwell within us! When we talk about the Holy Spirit it should be with reverence and honor. He wants to come within us to develop us as a temple of God, a place where God's presence can be housed. With this blessing there comes a warning. If you desecrate your body that has been made the temple of God, then you are desecrating God.

Our prayer should be, " Holy Spirit may I never grieve you, or hurt you, or lie to you, but may I respect you as God." He is the Blessed Holy Spirit. To understand Deity, we need to understand the Trinity. Some accuse us of believing that we have three Gods: God the Father, God the Son and God the Holy Spirit. This is not the truth. Our Christian theology is Bible based. " Within the nature of the One God there are three eternal persons, Father, Son and Holy Spirit." Here is the wonder of wonders. When I speak in my prayer language, which is the evidence of the Holy Spirit, I speak to God. Paul said to the Corinthian Church . " *I thank my God, I speak with tongues more than ye all"* (1 CORINTHIANS 14:18). Paul clearly indicates that when you speak in your prayer language you are speaking to God. He made that clear in his teaching of the gifts of the Spirit. " *If any man speak in an unknown tongue, let it be by two, or at the most by three, and that by course; and let one interpret. But if there be no interpreter, let him keep silence in the church; and let him speak to himself, and to God* " (1 CORINTHIANS 14:17-18). I can have a personal relationship with our Almighty God, through the person of the Holy Spirit. We need to express to the Holy Spirit a thankful heart for making it possible that our lives will be a clean vessel for God, so that we will be empowered to do His will. We know that the Holy Spirit is God because of His attributes.

The Holy Spirit is omnipresent. He is not just a mere expression, or mentioned when a benediction is given. He is present in all places, at all times. *"Where can I go from your Spirit? Where can I flee from your presence? If I go up to the heavens,* *you are there; if I make my bed in the depths, you are there. If I rise on the wings of the dawn, if I settle on the far side of the sea, even there your hand will guide me, your right hand will hold me fast"* (PSALM 139:7-10 NIV). You cannot hide from the Holy Spirit. No city is too large, no country too remote, He knows exactly how to convict you of your lifestyle.

The Holy Spirit is omniscient. For the Holy Spirit will give us infinite awareness, understanding, knowledge and insight. *"Who hath directed the Spirit of the LORD, or being his counselor hath taught him? With whom took he counsel, and who instructed him, and taught him in the path of judgment, and taught him knowledge, and showed to him the way of understanding?"* (ISAIAH 40:13-14). The obvious answer is that no one taught Him. He is God. He knows all about us He reads the intent of our heart.

The Holy Spirit is omnipotent. He has unlimited power, authority and influence. The angel said to Mary the Mother of Jesus. *" The Holy Spirit shall come upon thee, and the power of the Highest shall overshadow thee: therefore also that holy thing which shall be born of thee shall be called the Son of God"* (LUKE 1:35). This Scripture indicates that in the Holy Spirit resides the power of God. The Word of God also indicates that the power of the Holy Spirit works miracles. *"Through mighty signs and wonders, by the power of the Spirit of God; so that from Jerusalem, and round about unto Illyricum, I have fully preached the gospel of Christ"* (ROMANS 15:19).

When you bear witness to the world that you are filled with the Spirit, you are saying loud and clear that you are filled with God. You have Deity in you because of the Holy Spirit. This is a great wonder!

6
SIX

The Sixth Wonder of the Holy Spirit
HE IS ACTIVE IN THIS PRESENT DAY

We are hearing reports that the activity of the Holy Spirit has ceased to exist since the close of the first century. People want us to believe that the power of the Holy Spirit, which demonstrated great results in the early church has diminished This cannot be true because there are millions of people throughout the world that have experienced the ministry of the Holy Spirit. Peter's message on the Day of Pentecost explains the length of the activity of the Holy Spirit. *" Then Peter said unto them, Repent, and be baptized every one of you in the name of Jesus Christ for the remission of sins, and ye shall receive the gift of the Holy Spirit. For the promise is unto you, and to your children, and to all that are afar off, even as many as the Lord our God shall call"* (ACTS 2:38-39). The term *" even as many as the Lord our God shall call"* includes us in this day.

What is the Holy Spirit doing today? Some say that He is doing nothing. They have wrapped Him up in tradition as a mere story of an unusual happening which they now celebrate once a year in June, which is called Pentecost Sunday. There are others however, who claim that He is engaged in an outpouring of miracles such as man has never seen. Then there are others who give the Holy Spirit credit for everything from the outcome of political elections, to the progress of the ecumenical movement. These things may be good but what did Jesus say would be the main work of the Holy Spirit? *" But when the Comforter is come, whom I will send unto you from the Father, even the Spirit of truth, which proceedeth from the Father, he shall testify of me,"* (JOHN 15:26). The key phrase is *" He shall testify of me."* This is the main work of the Holy Spirit. He will reveal the life and power of the Son of God so that we can say with assurance, *" I can do all things through Christ which strengtheneth me."* (PHILIPPIANS 4:13). The Holy Spirit guides me to be a follower of Jesus, so that I can grow in grace and in the knowledge of Christ. (2 PETER 3:18).

The Holy Spirit is a teacher. Jesus said: *" But the Comforter, which is the Holy Spirit, whom the Father will send in my name, he shall teach you all things, and bring all things to your remembrance, whatsoever I have said unto you"* (JOHN 14:26). What does He teach? He teaches truth about the Lord Jesus Christ. He guides all believers into truths that they need to know for their spiritual development. His special ministry to the apostles was to bring to their memory all the things that the Lord Jesus had spoken to them. Jesus indicated that there were things He wanted to reveal to them but they would not understand at that present time. It will be the Holy Spirit that will reveal the great mysteries of God. Jesus said, *" I have yet many things to say unto you, but ye cannot bear them now. Howbeit when he, the Spirit of truth, is come, he will guide you into all truth: for he shall not speak of himself; but whatsoever he shall hear, that shall he speak: and he will show you things to come"* (JOHN 16:12-13). If we can understand that His special ministry to us, as it was to the apostles, is to bring to our memory the things that Jesus spoke and wanted to say. Our prayer should be: " Holy Spirit reveal to me the things that Jesus desires for me to know and for me to do." We should never forget that the activity of the Holy Spirit is to reveal a revelation to us concerning the manifestations of God and the workings of Jesus Christ on the earth. We need to be careful for any ministry or person who claims to be led by the Holy Spirit must not act contrary to the Word of God.

There is another purpose for the Holy Spirit and it is to convict unbelievers of their lifestyle. *"And when He is come, he will reprove the world of sin, and of righteousness, and of judgment"* (JOHN 16:8).

The Holy Spirit Convicts In Three Areas.

Jesus said that the Holy Spirit's special ministry to the unsaved is **to reveal sin,** *"Because they believe not on me"* (JOHN 16:9). The greatest sin a person can commit is the " sin of unbelief," and the worst kind of unbelief is to reject the Lord Jesus Christ. There is a " spiritual blindness" that prevails of the human heart. Thank God the Holy Spirit has come to give us light.

Jesus also said that the Holy Spirit will **expose human unrighteousness.** *"Because I go to my Father and ye see me no more"* (JOHN 16:10). Jesus is God's standard of perfect righteousness. Since His ascension into heaven, it has been the work of the Holy Spirit to convict men of their righteousness that is compared to filthy rags (SEE ISAIAH 64:6). He does it primarily by pointing men to Christ. The Bible clearly reveals that Christian faith is " *not by works of righteousness which we have done, but according to his mercy He saved us, by the washing of regeneration, and renewing of the Holy Spirit"* (TITUS 3:5).

Jesus encourages us that the Holy Spirit will **warn against future judgment.** " *Because the prince of this world is judged"*

(JOHN 16:11). Satan is already judged. Under the guidance and direction of the Holy Spirit we are led out of our condemnation because Jesus paid the price for our redemption. Here is the wonder of the Holy Spirit. He is saying, I have an activity that you cannot exhaust, the sky is the limit, and the impossible can be made possible.

The Seventh Wonder of the Holy Spirit
HIS GENEROSITY IS TO ALL BELIEVERS

The Bible plainly teaches that each believer is the dwelling place of the Spirit of God. This is a staggering truth. Think of it! God Himself lives in every Christian. The Bible expresses it this way: *"In the last day, that great day of the feast, Jesus stood and cried, saying, If any man thirst, let him come unto me, and drink. Whoever believes in me, as the Scripture has said, streams of living water will flow from within him. By this he meant the Spirit, whom those who believed in him were later to receive. Up to that time the Spirit had not been given, since Jesus had not yet been glorified"* (JOHN 7:37-39 NIV). The " flow" is from your inner most being, which tells me that it is a personal experience within the believer. The Holy Spirit will give strength for our weakness. He guides us in understanding God's Word and will. He helps us to pray. He empowers us to serve. He comforts us in our sorrow. He does it all by pointing us to Jesus Christ. This is the generosity of the Holy Spirit.

Jesus chose the word " baptism" in Acts 1:5 . *"For John truly baptized with water; but ye shall be baptized with the Holy Ghost not many days hence."* When one is baptized with the Holy Spirit as

Jesus taught in John 7:38,39 *"out of his belly shall flow rivers of living water. But this spake he of the Spirit, which they that believe on him should receive."* God's generosity in giving the fullness of the Holy Spirit is linked to a promise. *"For the promise is unto you, and to your children, and to all that are afar off, even as many as the Lord our God shall call"* (ACTS 2:39) . There is a continuous blessing and generosity because of a continuous flow.

If you are familiar with a water tower, it is filled with water and if you want water from the tank to your home there must be pipes and a faucet. Using this illustration, we are the faucet, and the Holy Spirit is the pipe that comes from God's great reservoir. We control the faucet. All of the economy and generosity of God are determined by how much of the Holy Spirit you release through your life. Why would someone want a piece of the rock when you can have the whole rock. Why would someone want a few drips from the living waters of God, when you can have a gusher flowing. The fullness of the Holy Spirit is not only a distinct and universal blessing of the new covenant, it is also a continuous blessing, to be continuously and increasingly appropriated. He gives, and He gives and He will not stop giving.

Here is a Scripture that deserves to be printed in gold. *" In the last day, that great day of the feast, Jesus stood and cried, saying, If any man thirst, let him come unto me, and drink"* (JOHN 7:37). Jesus was attending the Feast of Tabernacles, a colorful ritual that was conducted out side the Temple. Hundreds of people would gather as the priest would pour water from a golden pitcher over a rock. The people would then sing: *" With joy shall ye draw water out of the wells of salvation"* (ISAIAH 12:3). It was at that moment as Jesus stood up before the people He saw an emptiness, He saw that something was missing. Jesus stood up and with a loud voice said, " If you really want living water, I would like to tell you how to draw living water from the wells of God's economy. Come to me and drink. Be filled with the Holy Spirit and out of your innermost being will flow rivers of living water."[paraphrased] This inner experience will not be centered in the Feast of Tabernacles, it will be centered in the Third Person of the Godhead, the Blessed Holy Spirit! Be filled with the Holy Spirit.

Here is the wonder of the Holy Spirit. It is what He desires to do in you. Look at what is to characterize the life of one who is filled with the Spirit. *He is a singing Christian* (EPHESIANS 5:19). He speaks in psalms and hymns and spiritual songs. His heart is filled with a melody of praise to God. *He is a satisfied Christian* (EPHESIANS 5:20). He is always giving thanks and has a grateful heart, even in the adverse circumstances of life. *He is a submissive Christian* (EPHESIANS 5:21). He is yielded to the Lord. He is concerned for the needs of others. He is not self-centered. Change Him from a symbol to a person within you and you will be a dynamic, witnessing, fruit-bearing Christian.